WHY THE EARTH QUAKES

also by MARIO SALVADORI

WHY BUILDINGS STAND UP

also by MATTHYS LEVY *and* MARIO SALVADORI

WHY BUILDINGS FALL DOWN

Why The Earth Quakes

MATTHYS LEVY *and* MARIO SALVADORI

Illustrations by MICHAEL LILLY

W·W·NORTON & COMPANY
New York London

The text of this book is composed in Aster
with the display set in Inverserif Light
Composition and manufacturing by The Maple-Vail Book Manufacturing Group.
Book design by Jacques Chazaud

Library of Congress Cataloging-in-Publication Data

Levy, Matthys.
Why the earth quakes : the story of earthquakes and volcanoes.
p. cm.
ISBN 0-393-03774-6
1. Volcanoes. 2. Earthquakes. I. Salvadori, Mario George. II. Title.
QE522.S23 1995   95-1021
551.2'1—dc20

ISBN 0-393-31527-4 pbk.

W. W. Norton & Company, Inc., 500 Fifth Avenue, New York, N.Y. 10110
W. W. Norton & Company Ltd., 10 Coptic Street, London WC1A 1PU

1 2 3 4 5 6 7 8 9 0

To the victims of
nature's fury

Contents

Preface

Mario's experience:

"Many years ago, when I was thirteen, within a few days, I lived through two earthquakes. Since then all my friends have asked me, 'How did it feel? Were you afraid? What did you do?' All unanswerable questions, because the experience of an earthquake is so unreal that it is hard to describe: we expect *anything* to move but *not* the earth! The first earthquake caught me as I was getting out of bed on a summer morning and I was so surprised to see the lamp hanging from the ceiling move back and forth and feel the floor shake under my feet that I had no time to be scared. I ran to my parents' bedroom and my father said: 'This is an earthquake, but it is not very strong. Let us all stand underneath the entrance door for a while.' So we did, and after about five seconds the earth stopped shaking. We felt relieved, but even five seconds, if you count them by saying, 'One thousand and one, one thousand and two, one thousand . . .', become an eternity. By the time the earth finally stopped shaking I *was* scared!

The second earthquake was quite a different story. I was asleep in bed when a roaring sound woke me up, a sound seeming to come from the bowels of the earth. Then the mirror shattered on the floor, the glass panes of the window cracked, chunks of plaster fell from the ceiling and the walls began to crack. This time I knew it was a real earthquake and I ran straight to the front door. My father said, 'I am afraid this is a strong one. Our house is well built, yet some of the walls did crack. There must be a lot of damage all around us. Seismologists will call it an earthquake of high intensity on the Mercalli scale.' I did not understand the meaning of these last words, but this time I counted up to twenty before the earthquake stopped, and in the afternoon the earth shook slightly, once or twice. This time I was scared from beginning to end."

Matthys' experience:

"I arrived in Tokyo in the spring of 1952 for rest and rehabilitation while serving in the army during the Korean War. We were billeted in a big old hotel in the center of town and my room was on the tenth floor. Of course, I had read about earthquakes and had always been fascinated by Frank Lloyd Wright's Imperial Hotel, which had survived the 1923 Tokyo earthquake, in Wright's words, 'exactly as I had designed it to behave'. I was sound asleep when, in the middle of the night, I was rudely awakened and brusquely thrown to the floor. My first thought was that my roommates were playing a joke on me, but as I opened my eyes and saw them either on the floor or weaving about as if drunk, and yelling unprintable expletives, I knew that I had been shaken out of bed by my first earthquake. It was all over in a fraction of a minute and, as I learned later, was considered a minor event by the residents of Tokyo. But it was unforgettable to me: I spent many subsequent sleepless nights wishing I were closer to the ground."

Ever since these events early in our lives, we have been fascinated by earthquakes. Where do they come from? How come the earth, strong enough to support a skyscraper, can shake and crack? How often do earthquakes occur? Will we ever be able to find out *when* and *where* they will occur? Are we all in danger of getting hurt, even killed, by an earthquake?

Since then, we have both learned the answers to many of these questions and have even designed buildings that have resisted earthquakes. It made us decide to write a book about earthquakes in the hope of making our readers less fearful of these scourges and, if the need ever arose, to know which precautions to take against them. But we had to include volcanoes in our story because their eruptions are so intimately related to the occurrence of earthquakes that we cannot understand one of these natural phenomena without understanding the other.

We hope you will have as much fun reading our book as we had writing it, but, from the bottom of our hearts, we hope it will never, ever become *useful* to you!

Introduction

From earliest antiquity, humanity has searched for a way to explain the mysterious and terrifying natural phenomena we call earthquakes and volcanic eruptions. Throughout the ages, the most influential minds of all cultures have attributed the strangest causes to such mysterious events. Among the Maori of New Zealand, Ruaumoko, the god of volcanoes and earthquakes, is said to have been accidentally pressed into the earth as his mother turned face downward while feeding him. According to the legend, he has been growling and spitting fire ever since (Fig. 1).

Volcanic eruptions have always been attributed to fires within the earth, but of what source? Almost twenty-four hundred years ago, Aristotle (384–322 B.C.), the master of Greek thought and wisdom, believed that mild earthquakes were caused by wind escaping from caves within the bowels of the earth and severe shocks by gales that found their way into great subterranean caverns. We do not know whether all Athenians believed him, but we do know that

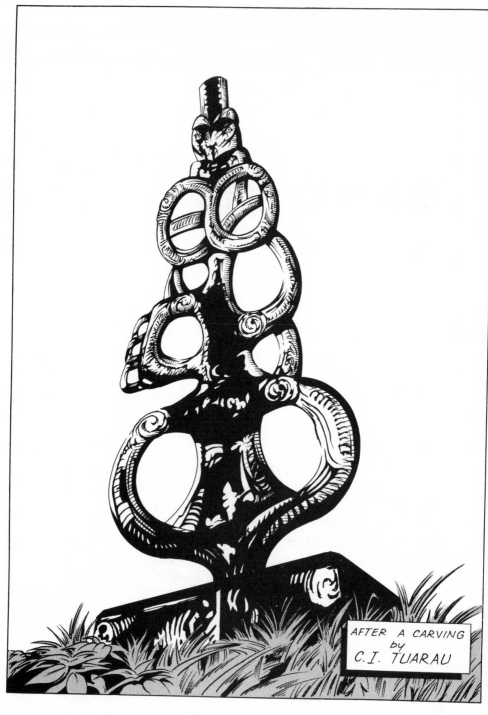

AFTER A CARVING by C.I. TUARAU

Fig. 1   Ruaumoko, the Maori God of Volcanoes and Earthquakes

any Roman expressing doubts about Aristotle's explanation was silenced by a short "Ipse dixit" ("He himself said so!"). Faith in Aristotle's statements was to last another two thousand years. Although at the time he could not look down from space and actually see the round shape of our planet as we now can, he observed two phenomena that supported his hypothesis that the earth was round: whenever the earth passed between the sun and the moon, in what we call a *lunar eclipse*, the earth's shadow on the moon was round (Fig. 2), and when he traveled southward, the North Star appeared lower in the sky (Fig. 3). Of course, he also mistakenly thought that the earth was the center of the universe, but from his point of view, as he stood on the earth and looked at the sky, this assumption seemed perfectly rational.

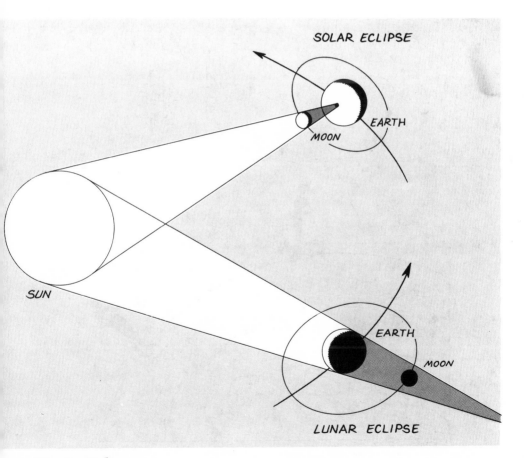

Fig. 2   Eclipses

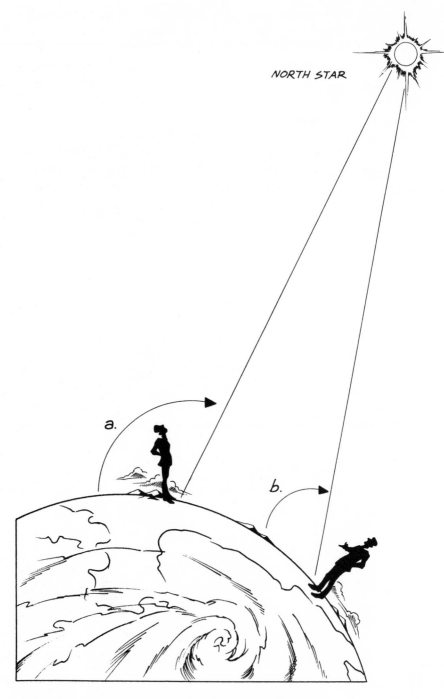

Fig. 3   Apparent Altitude of the North Star

Fig. 4   The Tortoise Supports the Earth

Even more exotic—even poetic—was the belief of some Native Americans that the earth was supported by a giant tortoise and trembled when the beast took a step (Fig. 4). We who live in the era of science have at long last begun to unravel the mystery surrounding the causes and consequences of natural events. We still do not entirely understand "how earthquakes work", and predicting *when* earthquakes may occur still baffles us, but we have been able to forecast quite accurately *where* they will happen and even how powerful they might be. Join us then on the fascinating trip that leads to our new understanding of earthquakes and volcanoes, a trip that started about twenty billion years ago with the creation of the universe, reached the birth of our solar system and of our earth (only five billion years ago) and briefly stops today before going into a future, certainly filled with ever more fascinating discoveries.

(The reader who would like to follow this story from its historical origin may wish first to read chapter 15 before visiting our earth to uncover its seismic mysteries.)

WHY THE EARTH QUAKES

1

The Cracked Egg

Thou unconcern'd canst hear the mighty crack:
Pit, box, and gall'ry in convulsion hurl'd,
Thou stand'st unshook amidst a bursting world.

—ALEXANDER POPE

Mother Earth

The **earth**,[1] when formed about five billion years ago, consisted mainly of very hot gases that, over time, cooled and became denser. As far as we know, the earth is the only grain of matter in the cosmos[2] that has an atmosphere conducive to life and, when unpolluted, is a joy to breathe.

If we could cut the earth in half like an apple (Fig. 1.1), we would notice its inner spherical *core* of solid nickel and iron (20 percent of the earth's radius), surrounded by a layer of molten nickel and iron (35 percent of the earth's radius). Around the earth's metallic center lies the *mantle* (45 percent of the earth's radius) of melted rock with a viscous, molasses-like but dense and

[1] Words and phrases that are highlighted in boldface type are defined in the glossary following Chapter 15.

[2] From the Greek *kosmos*, for universe.

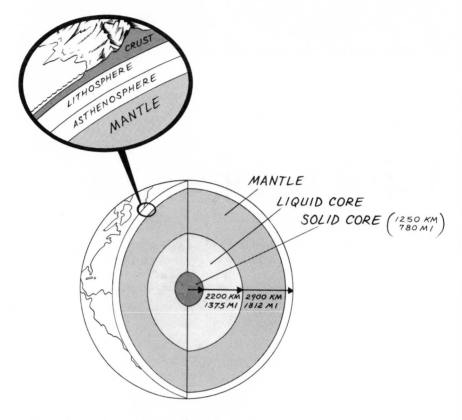

Fig. 1.1   The Structure of the Earth

fairly rigid inner layer, the *asthenosphere*,[3] a more plastic layer,
and the *lithosphere*,[4] the hard outermost layer 100 km (62 mi.)[5]
thick. The topmost outer layer of the lithosphere is the *crust*, on an
average 30 km (19 mi.) thick under the continents and only 5 km
(3 mi.) thick under the oceans. (The boundary between the crust
and the mantle, the *Moho discontinuity*, is named after the Yugo-
slav seismologist Andrija Mohorovičić, who first discovered it by
studying the path of seismic waves [see p. 68]).

[3] From the Greek *asthenés*, for weak.

[4] From the Greek *lithos*, for stone.

[5] Measurements are given primarily in the SI system and parenthetically in
English units.

Tectonics, or
Where Earthquakes Come From

As the earth cooled, the lithosphere cracked like an eggshell and
split into seven large and twelve small floating islands with ragged
edges: the *tectonic plates*,[6] that move continuously over the viscous
mantle, rubbing and / or pushing against each other and even trying to mount one over the other (Fig. 1.2). The majority of earthquakes originate along the plate edges, which often also define the
coastlines of the continents. Over two thousand years ago, the

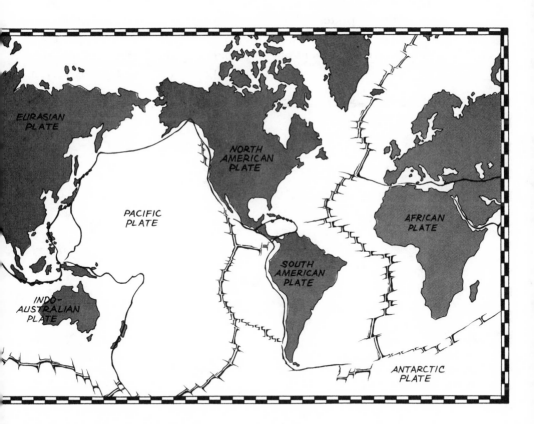

Fig. 1.2   Tectonic Plates

[6] From the Greek *tektōn*, for carpenter, the man who cuts wood planks.

Greek geographer, historian and philosopher Strabo first observed that active seismic regions lay along coastal bands, then called "rings of fire", probably because earthquakes were attributed to volcanic activity (Fig. 1.3). Geologists and seismologists have recently identified two primary bands of seismic activity: the *circum-pacific* belt that rings the coasts of the Pacific Ocean, and the *alpide* belt, along the southern boundary of the Eurasian plate, that cuts from the Atlantic Ocean across the Mediterranean into Asia. Since the ancient western world was centered in the Mediterranean region along the alpide belt, it is easy to understand why Greek philosophers sought to explain the origin of earthquakes.

Tectonic plates, which on an average move only about 50 mm (2 in.) per year over the mantle, are driven by *convection currents*, the upward movement of heated particles rising through the mantle from the earth's molten core (Fig. 1.4). As one plate is driven

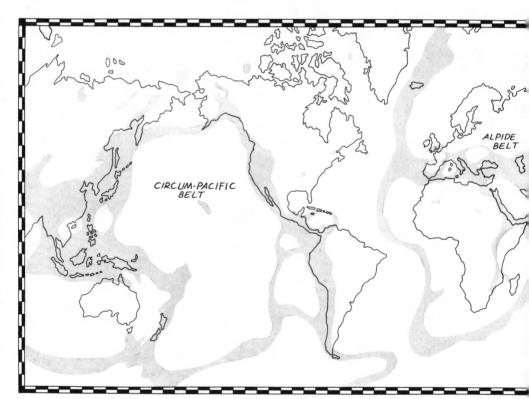

Fig. 1.3   Earthquake Zones: The Rings of Fire

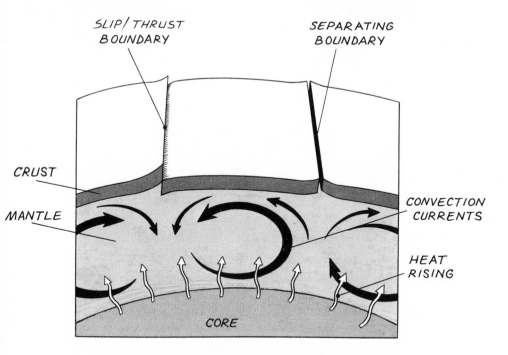

Fig. 1.4   Convection Currents Drive Tectonic Plates

against its neighbor, it will either try to (and eventually will) *slip* along the boundary (Fig. 1.5a), slide down (Fig. 1.5b) or one plate (usually a continental plate) will be *thrust* up over the other (usually an oceanic plate) (Fig. 1.5c). These three actions occur at *convergent* or *subduction boundaries,* regions where volcanic eruptions and earthquakes most often occur. At those tectonic boundaries where plates tend instead to *separate,* mainly along the mid-ocean *rifts* and *ridges* (Fig. 1.6), red-hot *magma* flows up from the mantle to fill the void left by the spreading plates and cools rapidly when meeting the frigid seawater, which also causes cracking in this new crustal material of *basalt.*[7] In this way, new material is constantly being added to the crust to make up for that which plunges back into the mantle and is remelted along a subduction boundary (Fig. 1.7).

The constant movement of the tectonic plates is resisted along their convergent boundaries by friction between their rough edges,

---

[7] An igneous rock formed from the molten magma and named after *basaltes,* a dark marble found in Ethiopia.

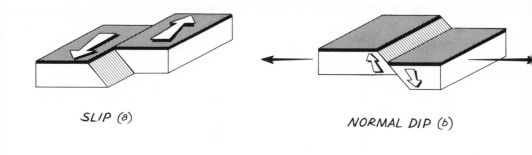

SLIP (a)

NORMAL DIP (b)

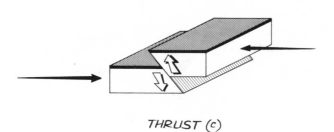

THRUST (c)

Fig. 1.5   Subduction Boundary Movements

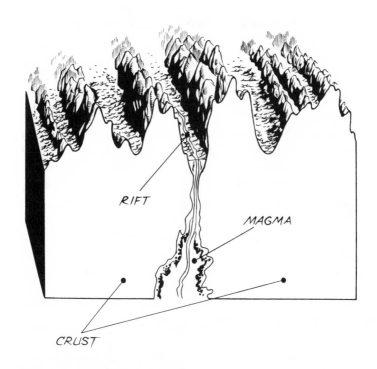

RIFT

MAGMA

CRUST

Fig. 1.6   Mid-Ocean Rift

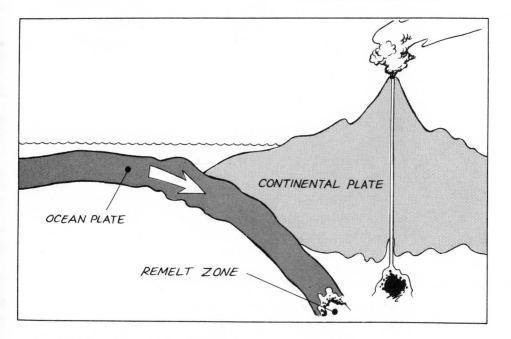

Fig. 1.7   Subduction Boundary

just as two bricks resist sliding when pushed one against the other. Over time this causes **strain,** which in turn causes **stress** to build up along the boundary, until a stress level is reached sufficient to overcome the plates' frictional resistance, causing a sudden slip (Fig. 1.8). This is how an earthquake occurs. Since tectonic plate theory is a mere forty years old, such a clear understanding is a recent, resounding triumph of seismology of which we may be proud.

Generally, major earthquakes occur along convergent boundaries on the continental coastlines, while minor earthquakes occur along mid-ocean separating boundaries. Approximately 70 percent of the continental earthquakes take place along the perimeter of the Pacific plate and 20 percent along the alpide belt, with the remaining 10 percent sprinkled around the globe.

Hot spots

Hot spots (Fig. 1.9) are places on the earth where columns of hot basaltic rock rise slowly from the mantle, puncturing the crust, forming volcanoes and causing earthquakes. When a tectonic plate

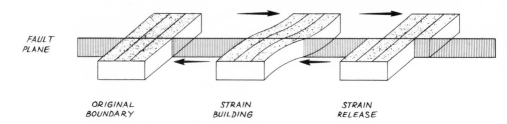

FAULT
PLANE

ORIGINAL   STRAIN   STRAIN
BOUNDARY   BUILDING   RELEASE

Fig. 1.8   Strain along a Fault Plane

lingers over a hot spot and then moves on, a volcanic island is formed. This tectonic plate movement will eventually result in a chain of volcanic islands like those in Hawaii or a *seamount*, an underwater ridge that rises above the level of the surrounding seafloor. The line of volcanoes that stretches from offshore Siberia to Kilauea is proof of tectonic motion. The Emperor Seamounts, a chain of underwater mountains in the northern Pacific, are ancient, extinct volcanoes that once were directly over the hot spot that is now below Hawaii. The mountains south of the Emperors are progressively younger and, in Hawaii, are active volcanoes.

Drifting Continents

Plate tectonic theory has provided the answer to a puzzle first mentioned by the English philosopher and scientist Francis Bacon (1561–1626): "Why do the facing shorelines of the continents look as if they could nestle into each other, the east coast of South America into the west coast of Africa, the 'bulge' of Africa into the Carribean and the eastern United States, and so on?" (Fig. 1.10). In 1912, the German geologist Alfred Wegener (1880–1930) suggested an explanation of this puzzle by proposing that about 200 million years ago the earth's crust consisted of only one supercontinent, Pangea, and one ocean, Panthalassa.[8] He justified his hypothesis by noting that similar rock formations, similar fossils and, particularly, the same kind of coal were found on both sides of the Atlantic Ocean in areas with boundaries that could nestle into each other if the continents were pushed together.

Through the study of plate tectonics, modern scientists have

---

[8] These words are from the Greek, *pan* meaning all, *gea* meaning earth, and *thalassa* meaning sea.

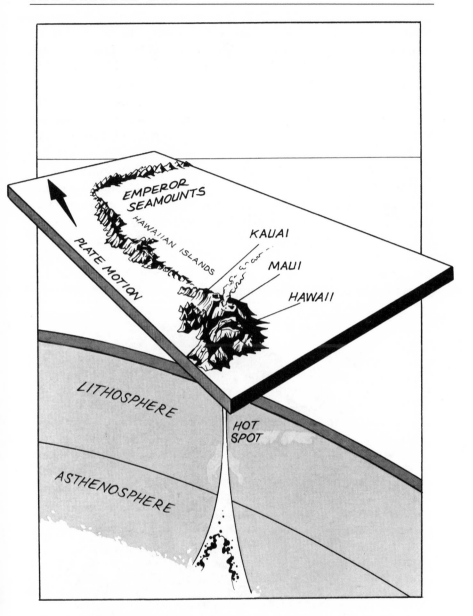

Fig. 1.9   Hot Spots

recently shown that the continents, and those tectonic plates whose edges extend beyond the continental boundaries seem to fit together like a gigantic jigsaw puzzle. It is now believed that Pangea broke up starting along a line that would become the mid-

Atlantic ridge (Fig. 1.11), when a separating boundary formed (starting from a preexisting fracture at the southern end of South America / Africa), then widened and opened the rift like a zipper at the rate of 100–200 mm (4–8 in.) per year. The *new* continents then moved apart for 160 million years and, by pushing and folding up the earth's crust, created many of the great mountain chains, like the Himalayas, that are found close to the edge of a convergent boundary.

According to this fantastic story, the Indian subcontinent actually started as an island that broke free of Madagascar and sprinted northward at about 200 mm (8 in.) per year toward Asia (Fig. 1.12); as India then crashed into and dipped under the massive Asian continent, huge masses of granite were pushed upward, raising the Himalayas. This movement continues today, as the Indo-Australian plate pushes northward at about 50 mm (2 in.) per year, further elevating the Himalayas.

Fig. 1.10   Pangea

Fig. 1.11   Drifting Continents

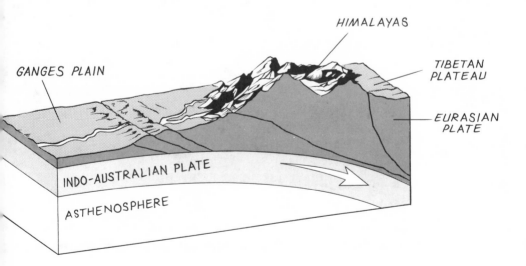

Fig. 1.12   Raising the Himalayas

Dinosaurs and other primitive reptiles roamed freely over these floating moving islands, sole witnesses to the evolving geography of the continents, but then, sadly, they faced extinction as the continents finally approached their present position on the surface of the earth (Fig. 1.13).

Wegener died in 1930 and his theory of continental drift languished while scientists argued over how the continents could possibly move. The first physical clue supporting his theory came from the discovery, twenty-five years later, of the *mid-ocean ridge*, a 64 000 km (40,000 mi.) long chain of suboceanic mountains that stretches along the bottom of the Atlantic Ocean, from north to south, and that curves around the southern end of Africa, with one branch bending north toward Arabia and another heading first east between Australia and Antarctica and then north along the west coast of the Americas (Fig. 1.14).

A deep zig-zag-shaped trench called the *great global rift*, which

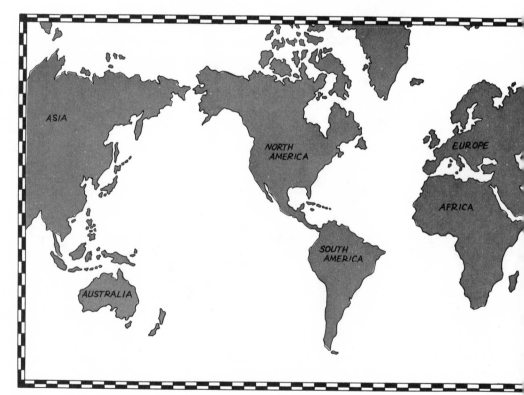

Fig. 1.13   The Continents

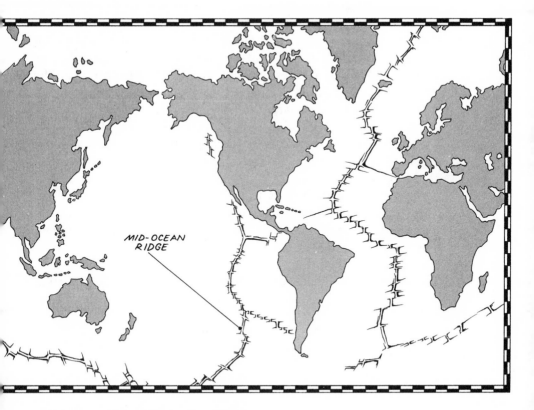

MID-OCEAN
RIDGE

Fig. 1.14   The Mid-Ocean Ridge

extends along the center of the whole length of the ridge, was discovered in the late 1950s. This discovery was the last piece of the
puzzle to convince scientists that the rift marked the edges of tectonic plates and that the separation taking place along these rifts
was driven by hot magma pushing upward, providing the engine
for Wegener's theory of drifting continents (although it was later
proved to be the tectonic plates and not just the continents
attached to them that were moving).

Wegener's theory has led earth scientists to even more
astounding discoveries. It is now believed that, in earth's history,
this division of the supercontinent into separate continents is
cyclically reversed and repeated every 440 million years: the continents split and drift apart for 200 million years, but then begin to
move back together again, eventually re-forming a supercontinent.
Scientists now assume we are at the end of one such cycle and that
the continents will soon begin to drift back together again. In a

mere 200 million years, they will re-form a supercontinent, making us at long last the inhabitants of "one world", only to begin the long cycle of separation and possibly senseless fights all over again, as if it were the destiny of humanity never to be at peace for more than a few million years!

It must now be apparent to the reader that our earth is a restless, constantly changing, living organism and not the steadfast "terra firma" we implicitly trust. For billions of years after the earth was formed, it was a seething, uninhabitable cauldron, a land of biblical fire and brimstone, a land where sheets of lava flowed up from its bowels to create our great plains and where vents pockmarked the surface with volcanoes. Even today, as if to remind us of our past, our more peaceful abode sometimes unexpectedly shakes with a violent tremor or blows the top off one of its gentle-looking hills.

Let us then first explore volcanoes, the crucibles of creation, which with their eruptions and explosions give us a hint of our earth's origins.

2

The Fire Goddess

Scorched mountain, boiling sea,
Heaven a furnace dire,
All the earth like lotus flower
Blooming first in fire.
Mankind all too quickly blown
Sped in flaming swirls
E'en as the ravening fire mist
Burned three thousand worlds—

——Kiken Ijichi

The Birth of a Volcano

On Saturday morning, February 20, 1943, Dionisio Pulido, a Taras-can Indian, was walking through his cornfield on the high plateau of Mexico's Sierra Madre Occidental when he felt an unusual warmth radiating from the earth below his feet. He was used to feeling the sun shining down, warming his body and bringing life to his crops, but to have the earth radiating heat? . . . Something was wrong. "I was with my wife, Pabla, and our ten-year-old son, Cresencio, who were shepherding our little lambs, . . . when I heard the ground in front of me snort and saw the smoke come out of the earth. I thought the world had caught fire!"

Dionisio and his neighbors had been warned by the local engineer that, because earthquakes had struck the area during the previous eighteen days, a volcano might arise near the village of Paricutín. The next day, Dionisio returned to the field with his oxen and donkeys and observed a most curious phenomenon: a crack in the earth in the middle of his field was spewing up a steaming mass of material and throwing up rocks onto a mound already 3 m (10 ft.) high. Within days, violent eruptions threw liquid lava into the air that, granulated by the expanding gases, hardened into ash as it fell back to earth. Glowing rocks ("bombs" to the volcanologists) were hurled skyward and fell back to earth on top of the ash, soon transforming the mound into the kind of hill that, because of its conical shape, the volcanologists call a *cone*. *Lava*, a molten, gas-charged, vitreous (glass-like) material, began to ooze from the crack in the ground, rising through the cone and spilling over its flanks. Two months later, in early April, against the backdrop of the heavy ash fall, lightning storms of great intensity struck the area: lightning flashed upward every two minutes or so with sharp crackles from the top of the ever-rising cone. All the while, lava kept flowing, at rates of up to 30 m / hr (100 ft. / hr.), preceeded each time by violent explosions: in five months the cone turned into a 310 meter (1,033 foot) high mountain completely obliterating the cornfield and the village (Fig. 2.1).

In a few months Dionesio and his neighbors had seen the spectacular birth of a volcanic mountain, the Paricutín volcano, which stopped erupting nine years later after reaching a height of 450 m (1,500 ft.). Volcanologists rushed to Paricutín from all over the world, to study this most unusual phenomenon.

The Origin of Volcanoes

A volcanic eruption is a blowout at a point of weakness in the earth's crust. It starts when a movement, often due to the shaking

Fig. 2.1   Birth of Paricutín

of an earthquake, opens a *fissure* or crack in the crust: a volcano is thus nothing but a vent in the earth's crust. *Magma* from the earth's mantle, which consists of molten rock, gases and mineral crystals, is held under tremendous pressure by the weight of the earth's crust bearing down on it: as the magma seeks to escape, it flows into the fissure and is squeezed out, releasing some of its gases, which, in turn, expand and widen the original crack (it was this type of action that caused the hot magma to shoot up to the surface at Paricutín). Sometimes, seawater enters the crack and, contacting the hot magma, suddenly expands into steam that widens the crack, resulting in an explosive decompression that spews solids and gases out of the mouth of the volcano. In all cases, a plug of rapidly solidifying magma eventually closes the crack and, as happens in a self-sealing tire, stops the eruption.

It has been estimated that in the last ten thousand years, twelve hundred active volcanoes have erupted, of which five hundred are still active today in areas populated by 500 million people. Of these, 60 percent are located around the rim of the Pacific Ocean, also known as the *ring of fire*. The similarity in the worldwide distribution of volcanoes and earthquakes over the earth's surface shows beyond any doubt that a relationship exists between them: volcanoes are found along the seismic circum-pacific and alpide belts (see p. 24) and on islands, such as Hawaii and Samoa, that lie along the mid-Pacific rift or, like the Azores and Iceland, along the mid-Atlantic rift (see p. 25). Volcanoes also occur in the Alps and the Himalayas, regions of mountain folding that produce lines of weakness in the earth's crust, and most often along tectonic boundaries, as in the case of Mount Paricutín. Points of particular weakness in the earth's crust, such as the crossing of two fractures at Mount Vesuvius, become the seats of major volcanoes and are most often located near coastlines.

The Legacy of Vesuvius

The Bay of Naples is the most romantic spot on the Italian peninsula. The enchanted tourist, sitting in the shadow of a Roman parasol pine tree, gazes at the graceful curve of the shore, bound by Cape Misano on the north and Sorrento on the south, and admires the blue waters of the Mediterranean, dotted by the charming islands of bright Capri, smoking Ischia and tiny Procida.

Turning around to view the gently sloping land, fertile with the vineyards famous for its Lacryma Christi (Tear of Christ) wine, the tourist sees the elegant cone of Mount Vesuvius emitting a faint plume of white smoke—an idyllic landscape worthy of an eighteenth-century painting. But hidden under this lovely land lies a fiery cauldron that has been called the *crucible of creation* and is the source of all this natural beauty (Fig. 2.2).

Mount Vesuvius, the only active volcano on the European mainland, has been active since early antiquity, erupting violently in A.D. 63, 79, 512 and 1631; it erupted six times in the 1700s, eight times in the 1800s and three times in the twentieth century, the last time in 1944. Our tourist becomes keenly aware of the danger posed by Vesuvius to Naples while on a visit to the ruins of Herculaneum and Pompeii, the most serendipitous of archeological discoveries that has been thus described by the renowned archeologist, Paul McKendrick:

Fig. 2.2   Vesuvius Seen from Naples

One day in 1711 a peasant digging a well on his property on the bay, five miles southeast of Naples, came upon a level of white and polychromic marbles, obviously ancient. This chance led to the discovery of what proved to be the buried town of Herculaneum, destroyed by the eruption of Vesuvius in A.D. 79. In 1748, workmen digging the Sarno Canal, nine miles farther along the bay, discovered bronzes and marbles on a site which an inscription, discovered fifteen years later, identified as Herculaneum's more famous sister city, Pompeii. Thus began the saga of an excavation which has told the modern world more about ancient life than any other in the long history of archeology. . . . In a few hours of a summer afternoon the eruption stopped the life of two flourishing little cities dead on its tracks.[1]

The remains of the two cities had been kept intact for sixteen centuries by the products of the A.D. 79 eruption of Mount Vesuvius: Herculaneum drowned in a torrent that submerged it below as much as 25 m (75 ft.) of mud and Pompeii, buried under almost 8 m (26 ft.) of ash and lapilli (small volcanic stones). Undamaged were temples and villas, frescoes and wall signs, trade tools and even the encapsulated bodies of the victims, many holding the valuables they went back to recover only to find certain death. The red and black signs on the street walls, readable today, give an insight into the only too human feelings of the Pompeiians about politics: "The sneak thieves support Vatia for the aedileship [the position of head of the building department]", reads one; "Vote for X, he won't squander the public funds", reads another. They could have been scribbled today on walls anywhere.

Among the dead at Vesuvius was the great naturalist Pliny the Elder, author of a thirty-seven-volume natural history, who was asphyxiated while investigating the eruption. Seeing the rising cloud over Mount Vesuvius, the elder Pliny, wishing to investigate the eruption at close hand, took a boat from his seaside home and sailed toward Stabiae, inviting his nephew, Pliny the Younger, to come along. Excusing himself from the dangerous excursion because "he had too much homework to do", the younger Pliny later described his uncle's tragic but stoic death in a letter to the historian Tacitus:

[1] *The Mute Stones Speak*, Paul McKendrick (New York: W. W. Norton, 1983).

And now cinders, which grew thicker and hotter the nearer
he approached, fell into the ships, then pumice-stones too,
with stones blackened, scorched, and cracked by fire, then
the sea ebbed suddenly from under them, while the shore
was blocked up by landslips from the mountains. . . .

After landing, the elder Pliny, to show his lack of fear in the face of
the deepening danger, went to a house to take a nap. When he was
awakened, he stepped outside:

For the house now tottered under repeated and violent con-
cussions, and seemed to rock to and fro as if torn from its
foundations. In the open air, on the other hand, they
dreaded the falling pumice-stones, light and porous though
they were; yet this, by comparison, seemed the lesser danger of the two; a conclusion which my uncle arrived at by
balancing reasons, and the others by balancing fears. . . .
Soon after, flames, and a strong smell of sulphur, which
was the forerunner of them, dispersed the rest of the company in flight; him they only aroused. He raised himself . . .
but instantly fell; some unusually gross vapour, as I conjec-
ture, having obstructed his breathing and blocked his
wind pipe . . .

He was found the next morning looking more like a man asleep
than one who died from the effects of a violent volcanic eruption.

Neapolitans would not dream of abandoning their beloved city
despite the fact that, if the A.D. 79 eruption were to occur today,
200,000 people might be killed instead of 16,000, and 500,000
homes would be destroyed instead of 20,000 (18,000 died in the
1631 eruption, although the inhabitants of the area had been
warned of the imminent danger by the increasing activity of the
volcano).

Neapolitans are indebted to their volcano for what the legends
of the natives of Java, one of the most volcanic islands on earth,
describe as "times of darkness followed by times when plants grow
at a great rate with unheard of yields." So, with every period of
quiescence following a major eruption, farmers return to Vesuvius,
drawn by the incredibly rich and fertile soil, and cultivate the land
further and further up the cone until one day the mountain wakes
up and pushes them back down . . . and the cycle starts all over
again.

Mount Pelée

The fire goddess of the Kilauea volcano, Pelé, stands guard over the three active volcanoes of the Hawaiian Islands: Kilauea, Mauna Loa and Hualalai (born in Tahiti, the goddess was chased away by her sister and settled in the crater of Kilauea). These are known as quietly active members of the volcano family because their gentle eruptions, which spill slow-moving lava, are more like a pot of soup boiling over than like the violent, explosive eruptions of Mount Vesuvius.

The earth is dotted with thousands of *extinct volcanoes*, identifiable by their characteristic shallow conical shape, like those in the Auvergne (France) and in Catalonia (Spain). But *dormant volcanoes* are the most feared because, while seeming to sleep, they are still alive, and can reveal their hidden explosive nature most unexpectedly.

Mount Pelée, on the northern end of the Caribbean island of Martinique (not to be confused with Pelé, the Hawaiian fire goddess), can be visited today by driving on a gently rising road up to the edge of a small crater lake set in a mild depression at the top of the hill. It is probably the view seen by Columbus, the island's European discoverer, in 1502, and by the early eighteenth-century settlers of the port city of Saint-Pierre at the base of the volcano. Such a sight may well have fooled these early visitors into believing that Pelée was an extinct rather than a dormant volcano, but it did not fool the later inhabitants of the island after small eruptions shook it in 1792 and 1851. A few uneventful years went by and then, in the early spring of 1902, like the sounds of a sleeping giant stirring, deep rumblings were heard from Pelée, followed by the venting of smoke and steam. Toward the end of April, the mountain began to explode, spitting out its crater lake.

Just prior to this minor explosion, an earthquake shook Quezaltenango in Guatemala, 3 400 km (2,100 mi.) away; on May 7 a major volcano erupted at Soufrière, on St. Vincent, only 160 km (100 mi.) away; and on May 10, the Izalco volcano erupted in El Salvador 3 100 km (1,938 mi.) away (Fig. 2.3). These four events, occuring within days of each other—fractions of seconds in geologic time—provide strong evidence of the interrelationship between earthquakes and volcanoes (especially when we realize that these locations are on opposite ends of the Caribbean plate).

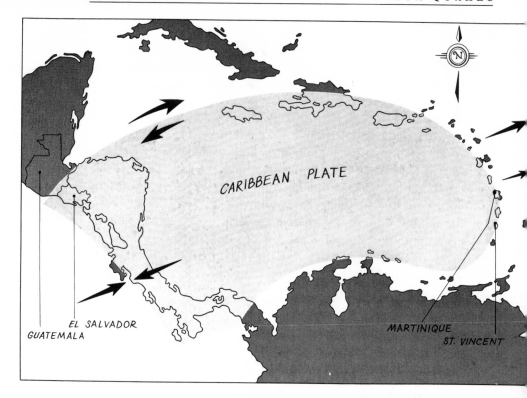

Fig. 2.3   Caribbean Volcanoes

The captain of a steamship tied up off the Saint-Pierre water-front reported that, shortly before 8 A.M. on May 8: "Violent detonations suddenly shook earth and sea. The mountain appeared to split open from summit to base, giving passage to a flashing flame which shot up into the air, and a gush of black clouds rushed down the slope of the mountain like a whirlwind. When it reached the base, it spread out like a fan toward the city which it plunged into darkness" (Fig. 2.4).

On another ship in the harbor, a governess from Barbados was dressing her young charges when: "The steward rushed past shouting 'close the cabin door, the volcano is coming!' We closed the door and at the same moment came a terrible explosion which nearly burst the eardrums . . . and seems to have blown the sky-light over our heads, and before we could raise ourselves hot moist ashes began to pour in on us. . . . A sense of suffocation came next . . . we were all covered in black lava and were ankle deep in hot mud."

As the darkness passed, what had been the city of Saint-Pierre was a mass of roaring flames. In a little over one minute, the intense black cloud, illuminated periodically by brilliant lightning, had engulfed the city and the ships lying in the harbor, killing nearly thirty thousand people, mostly by suffocation and heat. Such flow from a volcanic eruption is called a *glowing avalanche* or a **pyroclastic flow** because of the great mass of red-hot, granulated lava, the so-called *volcanic sand*, that rushes down the face of the volcano like an avalanche of glowing black snow. It is similar to the mixture of ash and condensed steam rising from the crater of Vesuvius which, turning into a flow of mud, buried Herculaneum.

While the frequency of eruptions at Mount Pelée decreased, the strength of the explosions did not. On August 30, an explosive eruption showered debris-laden steam from the crater on the survivors of the first eruption, taking two thousand additional victims. Smaller, simultaneous eruptions at both Mount Pelée and Soufrière continued well into the fall of 1902, joined by that of the Masaya volcano in faraway Nicaragua. At long last, the eruptions

Fig. 2.4   Mount Pelée Erupts

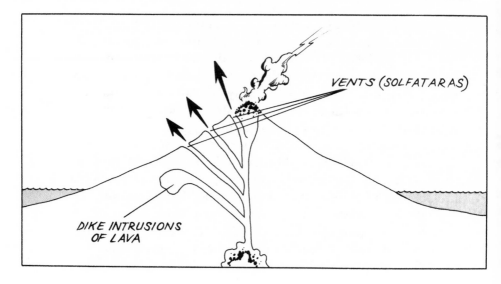

VENTS (SOLFATARAS)

DIKE INTRUSIONS
OF LAVA

Fig. 2.5   Solfataras

ceased after months of terror, but, as with all dormant volcanoes, the stage may potentially be set for new, unpredictable catastrophes.

The close relationship between volcanic and earthquake activities, exemplified by the events of 1902 in the Caribbean, can be explained by realizing that all volcanic cones, which eventually become circular cones, begin as linear rifts. In fact, all volcanic cones have lateral fractures along a vertical plane that is revealed on their surface as, for instance, the chain of craters at Kilauea, Hawaii, or the series of vents at Mount Pelée, (called *solfataras*, because of their sulfurous gas emissions) (Fig. 2.5). Whether they are cause or effect, movements of the earth's crust occur coincidentally during earthquakes and in volcanic eruptions. A volcano seeks to spill its deadly fruit at points of weakness along a rift, as vividly illustrated by the birth of Paricutín, when, two days after the first eruption, lava appeared not from the cone, but from a fissure 300 m (1,000 ft.) north of it. The hundreds of old volcanic cones dotting the plateau within 120 km (75 mi.) of Paricutín provide additional visible proof of the birth of volcanoes and their origins in earthquake country.

3

The Mountains
of Vulcan[*]

> The great rocks of the mountains will
> throw out fire; so that they will burn the
> timber of many vast forests, and many
> beasts, both wild and tame.
>
> —LEONARDO DA VINCI

Dormant Giants

Earthquakes are often the first signs of the reawakening of a dormant volcano and are sometimes accompanied by the venting of
steam and ash. Such events stimulate volcanologists to monitor
these activities with **seismographs** to measure earth movements,
inclinometers to detect changes in the slope of the land, thermom-
eters to monitor the earth's temperature and gauges to sniff the air
for traces of sulfur dioxide (smelling of rotten eggs) and other gases
that may be seeping out of the earth. Such careful monitoring of
pre-eruption activities was responsible for the evacuations that
saved thousands of lives when, in 1991, Mount Pinatubo erupted
in the Philippines. On April 2 of that year, the 1 259 m (5,770 ft.)

[*] The Roman god of fire.

high mountain on the island of Luzon, 87 km (55 mi.) from the almost seven million inhabitants of Manila, slowly awoke from a six-hundred year slumber. Steam and ash first blew out from vents on its sides, and then the lava dome on its top began to puff up. On June 11, when the dome had doubled in size, an evacuation was ordered on the basis of the volcanologists' prediction of an imminent eruption, a prediction fulfilled four days later when the mountain spewed out stones and lava and blasted ash and gases 40 km (25 mi.) into the atmosphere and invaded the land with devastating **pyroclastic flows.** Thousands of men, women and children abandoned their homes and reached safety by bus, horse-drawn carriage and on foot. . . . Thousands of them were still stranded in camps near the capital two years later.

The eruption is now believed to have been triggered by a series of earthquakes, starting with a magnitude 7.7[1] event in 1990 that allowed magma to penetrate a cavity under the volcano.

One may wonder at the relatively minor number of deaths—nine hundred, none among U.S. troops—in such a major catastrophe, but two reasons for this relatively small-scale death toll are obvious: the sparsely populated region of small villages around the volcano and the relatively lengthy distance of the volcano from the populous capital. Even though the rain of ash changed day into night and made breathing difficult in Manila, it did not cause death there. But the fundamental cause of the small number of deaths must be found in the dedication and exceptionally good judgment of the volcanologists and the timely and effective precautions taken by the authorities.

One glorious reminder of this dramatic event survived for months: Along a wide band straddling the equator (Fig. 3.1), as the evening sun slowly set below the horizon, the skies revealed their expected pink hue evanescing into blue, but instead of rapidly darkening, there first appeared a red, then a violet, and finally a phosphorescent purple color that lasted an unusually long time. This superb show of natural beauty, seen as far as the United States, was due to the volcanic particles of Pinatubo blown by the wind around the world, a strangely beautiful fruit of the brutal forces of nature.

[1] See Chapter 6 for the definition of magnitude.

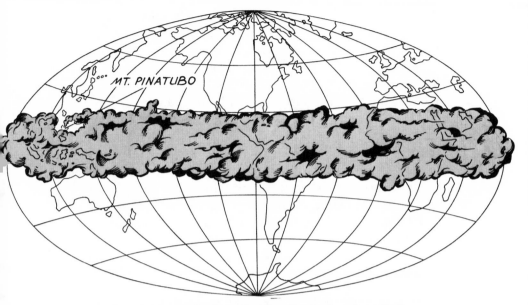

Fig. 3.1 Mount Pinatubo's Ash Cloud Circles the Globe

Mount St. Helens

The Cascade range of mountains extends along a 240 km (150 mi.) line, inland and parallel to the Pacific Coast of Canada and the United States, from British Columbia in the north to northern California in the south. It is a range dotted with volcanic peaks, of which Mount Lassen at the southern end, 3 187 m (10,457 ft.) high, last erupted in 1917; the northernmost peak, Mount Baker, 3 285 m (10,778 ft.) high, last erupted in 1843 and Mount St. Helens, a 2 549 m (8,364 ft.) high peak near the southwest corner of Washington State, had last erupted in 1842 and remained active until 1857 (Fig. 3.2). According to an Indian legend, an ugly woman named Loo-wit was transformed into the beautiful snow-capped peak of Mount St. Helens because, when war broke out between Indian tribes, it erupted, providing fire to the Great Spirit. Called Mount St. Helens when first seen by frontiersmen in the nineteenth century, the mountain had a regular conical form, suggesting it was a young volcano. It was then over 390 m (1,300 ft) higher than it is today, and even after the eruption of 1842, the only visible scars on its otherwise smooth surface were a lava track cutting through the forested land northward for 32 km (20 mi.) and remnants of *blowholes,* where heat had turned subterranean water into steam, that jetted through the hot viscous lava.

In early 1980 Harry R. Truman (no kin to the former president) looked out from the window in his lodge on the shore of Spirit Lake at the alpine landscape with its thick evergreen forests over a blanket of spring snow, exclaiming to his visitors, "what a beautiful country we got, boys." Truman, a feisty 84-year-old widower, lived alone but for the company of more than a dozen cats, in the lodge 8 km (5 mi.) northwest of the peak of Mount St. Helens. He was not surprised when the mountain awoke on March 25 of that year, after a 123-year-long sleep, sending out shaky precursors (earthquakes of a magnitude up to 4), and volcanologists predicted that an eruption was about to take place. After all, Truman said "I talk to the mountain, the mountain talks to me. I am part of the mountain, the mountain is part of me."

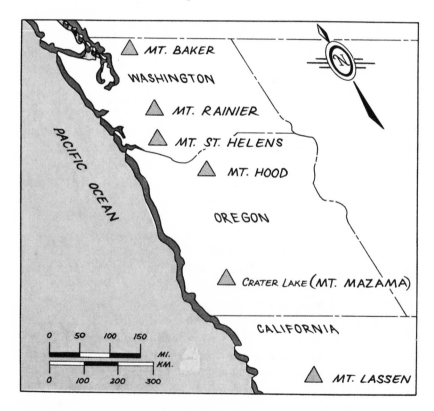

Fig. 3.2   Cascade Mountains

Three days later, Mount St. Helens started blowing off steam, spewing out ash and small boulders and propelling mud down its faces, while the heat rising from the volcano melted the snowpack. By March 30, a second crater was seen on the summit by Bob Christiansen of the U.S. Geological Survey (USGS), one of the many geologists and photographers drawn to the unfolding drama of the first stirrings of a long dormant volcano. There were more eruptions the next day, but officials stated there was no cause for alarm, since both Mount Hood and Mount Rainier (the highest of the range, at 4 352 m [14,410 ft.]) nearby in Washington State, had been exhaling vapors for years without a major eruption.

Nevertheless, observers noted that a disturbing bulge was developing on the north side of the mountain. Barely visible at first, it had grown at an impressive rate and by April 20 protruded 75 m (250 ft.) from the side of the mountain, while continuing to grow 1.5 m (5 ft.) a day (Fig. 3.3a). By this time, the two craters had merged into a single wide bowl, 255 m (850 ft.) deep and 510 m (1,700 ft.) wide, that on May 8, was deeply scarred when the 123-year-old volcanic plug blasted out of the throat of the crater. An increase in the frequency and magnitude of earthquakes in the following days (some up to magnitude 5), signaled a quickening of the volcano's activity. The north face, now puffed out menacingly more than 90 m (300 ft.), led Donal Mullineaux of the USGS to predict that a lava eruption could be triggered within a week by the increasing gravitational pull of the sun and moon. Residents were advised to leave the area, but Harry Truman stubbornly refused to move, insisting that "If I got out of here, I wouldn't live a damned day."

Sunday, May 18, dawned as a clear, sunny day with almost unlimited visibility. In the early afternoon, a magnitude 5.1 earthquake shook the mountain. Then at 4:30 P.M. the weakened rocks of its north face cracked, causing a huge landslide of almost 2 cubic kilometers [0.5 cu. mi.] of material (Fig. 3.3b). Through the opened fissures the pressure of the carbon dioxide and water vapor, which had been building up over the last few months, was suddenly released. An incredible blast, later compared to the explosion of a six-megaton bomb,[2] blew out the *entire* north face of the mountain, hurling ash, boulders and ice skyward (Fig. 3.3c). The force of the

---

[2] Such a nuclear bomb explosion has an energy equivalent to more than five hundred times the energy of the first nuclear bomb dropped on Hiroshima.

blast scraped clean the northern landscape, downing every tree within 27 km (16.8 mi.) of the summit (Fig. 3.3d).

Kran Kilpatrick of the U.S. Forest Service was planting trees 5 km (3.1 mi.) down the south side of the mountain. "There was no sound to it, not a sound—it was like a silent movie, and we were all in it. First the ash cloud shot out to the east, then to the west, then some lighter stuff started shooting straight up. At the same time, the ash curtain started coming right down the south slope toward us. I could see boulders [estimated to have been up to 12 m (40 ft.) in diameter] being hurled out of the leading edge, and then being swept up again in the advancing cloud." Then came the glowing ash cloud (the pyroclastic flow) rolling down the north face of the mountain at 320 km/hr (200 mph), igniting and snuffing out everything in its path. Ash and steam vented from every pore of the volcano, creating a billowing cauliflower cloud that spread out over 500 km² (195 sq. mi.). It suffocated Harry Truman and buried his lodge in a mountain of ash and debris from the landslide, which also raised the level of Spirit Lake by almost 60 m (200 ft.). Mud flowed down the mountain, filling the north fork of the Toutle River valley with 90 m (300 ft.) of ash, debris, tree trunks and glacial ice. The previously multicolored landscape that Harry had so admired was reduced to shades of gray, and everywhere, trees, stripped naked by the heat and force of the blast, were scattered like straw blown by the wind.

Because the area around Mount St. Helens was sparsely inhabited, a modest total of sixty-one people perished in the onslaught of ash and mud, many the very same geologists and photographers who were on the mountain to track the progress of the physical changes taking place prior to the eruption. (Caught by a sudden cataclysm, many have lost their lives in the service of science or out of pure curiosity: the French vulcanologists Maurice and Katia Krafft and the American Harry Elicken died in Japan in 1991, together with a group of thirty-seven Japanese journalists, cab drivers and farmers, in a cloud of heat from an eruption of Mount Unzen. Similarly, in 1993, a sudden eruption caught unawares a group of scientists collecting samples of gas in the crater of the Galeras volcano in Colombia, killing six of them.) But, at Mount St. Helens as at Pinatubo, the numerous warnings and the forced evacuations undoubtedly succeeded in keeping down the loss of life. Yet the mountain had not finished expressing itself on that fateful day: it erupted until the end of the year and, before stop-

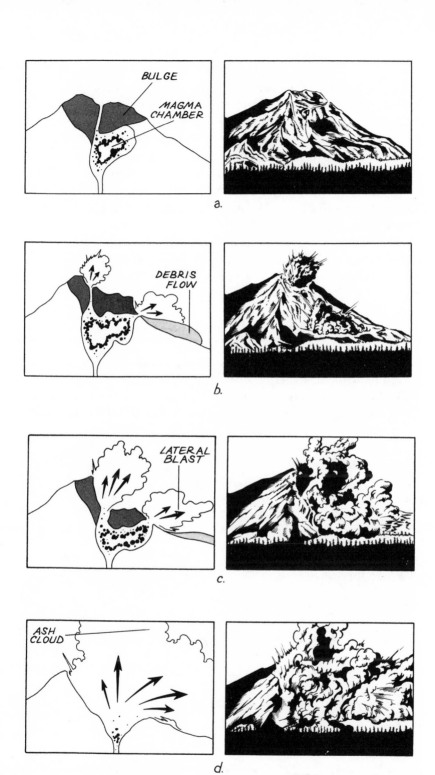

Fig. 3.3   Mount St. Helens Erupts

ping, the lava dome in the crater rose over 50 m (167 ft.). Since
Mount St. Helens is the most active and explosive volcano in the
United States and has erupted regularly every one hundred to five
hundred years, it may soon erupt again. But the possibility for the
repeat of a long-forgotten eruption in the Cascade Range is even
more frightening: about six thousand years ago, an explosion
twenty times more powerful than the 1980 eruption of Mount St.
Helens destroyed Mount Mazama, creating the tranquil beauty of
Crater Lake (Fig. 3.4).

Nature heals her wounds, and, ten years after the eruption, a
visitor to Spirit Lake once again sees the sprouting verdant paradise that attracted Harry Truman.

The Death of Krakatau

Two recent eruptions in Indonesia have come close to matching
the cataclysmic destruction of Mount Mazama in the Cascades.

Indonesia consists of an arc of languid tropical islands straddling the equator southeast of the Asian mainland in the Pacific

MT. MAZAMA·MT. ST. HELENS·VESUVIUS · MT. ST. HELENS· TAMBORA ·   KRAKATAU ·MT. ST.
4600 B.C.   1900 B.C.   A.D. 79   1500   1815   1883   19
(CRATER LAKE,)
(OREGON)

-NUMBERS INDICATE RELATIVE SIZE OF ASH CLOUDS-

Fig. 3.4   Relative Size of Volcanic Ash Clouds

Ocean. The largest island, Sumatra, lies off the coast of Malaysia and Singapore. To the east are the main island of Java and the hauntingly beautiful tropical island of Bali, with its gleaming white beaches, lush green terraced fields of rice, *and*, like the rest of the chain of islands, live volcanoes (Fig. 3.5). Once known as the Dutch East Indies, the panoramic islands of Indonesia make up one of the most active volcanic areas on earth, with as many as eighty active volcanoes hiding their explosive nature beneath a green blanket of dense vegetation. Since the 1600s they have suffered fifty-eight eruptions, including the great 1815 eruption on the island of Sumbawa, east of Java, that spewed out so much ash into the atmosphere that 1816 was called "the year without summer": crops failed in Europe, it snowed in New England in June and frost covered the southern United States on the fourth of July. But no more dramatic eruption took place there than that popularly known as Krakatau.

In the Sunda Strait between Java and Sumatra lies Krakatau, an island once the size of Manhattan, dominated in ancient times by a 2 000 m (6,900 ft.) high volcanic mountain (Fig. 3.6). Over the centuries, ancient Krakatau destroyed itself, most likely by violent eruptions, leaving four separate islands in a sea basin. On May 29,

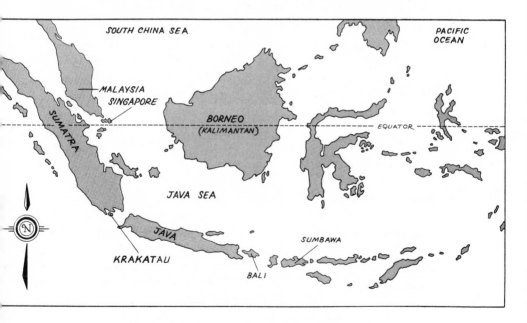

Fig. 3.5   Krakatau and the Indonesian Islands

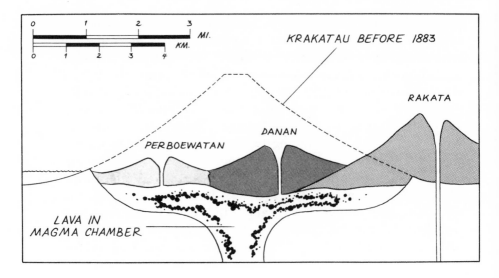

Fig. 3.6   Ancient Krakatau

1883, after two hundred years of dormancy, blasts of steam, laden with ash, began to spurt simultaneously from three vents—Perboewatan, Danan and Rakata (which were joined, through a fissure in the ground at the center of Krakatau island, to a subterranean crater, the root of the ancient volcano). A churning cauliflower-like cloud rose 11 km (7 mi.) in the sky and hung over the islands for three months, covering the ground with as much as 500 mm (20 in.) of ash. Sightseers from neighboring islands came by steamer to view the spectacular phenomenon and even dared to climb and look into the cones of the volcanoes amid the threatening rumblings from the earth. Though frightened by the constant roaring and the jets of steam rising skyward, none of these visitors seemed to be alarmed until, on the afternoon of August 26, a sudden explosion took place. Further explosions followed until 10:02 the next morning, when one of the most gigantic blasts ever recorded blew off two of the three cones, collapsing one of the islands into the sea and raising a black cloud of ash-laden steam 70 km (44 mi.) into the sky. The roar of the explosion was so powerful that it shattered the eardrums of sailors 40 km (25 mi.) away and was heard in Australia, 3 000 km (2,000 mi.) to the south. Ash from the volcanic cloud fell over an area of 827,000 km² (323,000 sq. mi.), twice that of the state of California. The cloud traveled around the world for over two years, reducing solar radiation and so causing cooler weather, anemic crops, and spectacular red sunsets, providing the world a true foretaste of a "nuclear winter". (A century earlier, the U.S. ambassador to France, Benjamin Franklin, had reported that a much smaller volcanic eruption in Iceland had caused the weather in Paris to be continually murky and cold even into the summer. Some observers noted that this change in the weather led to unrest in the populace and may have been a contributory cause of the French Revolution!)

The first explosion at Krakatau caused the disappearance of the Perboewatan volcano, allowing the ocean waters to rush into the newly created void (Fig. 3.7). Having filled the void, the ocean retreated in a giant wave, a **tsunami,** which radiated from the volcano at the incredible speed of 720 km / hr (450 mi. / hr.). Starting as a wave 30 m (100 ft.) high, it engulfed the town of Merak, 48 km (30 mi.) away, killing thousands of helpless people; it continued traveling across the Pacific Ocean, slowly dissipating its enormous energy, and ended by meekly slapping at the coast of Panama. A second explosion, an hour after the first, destroyed the Danan volcano, leaving only a submarine ridge at the bottom of the sea and generating a second tsunami that radiated outward, engulfing everything in its path. After all the explosions subsided, thirty-six thousand people were dead and only Rakata, the south end of Krakatau island, 800 m (2,700 ft.) high, stood as a reminder of the once proud 2 000 m (6,900 ft.) high Krakatau volcano (Fig. 3.8).

Fig. 3.7   A Tsunami Is Born

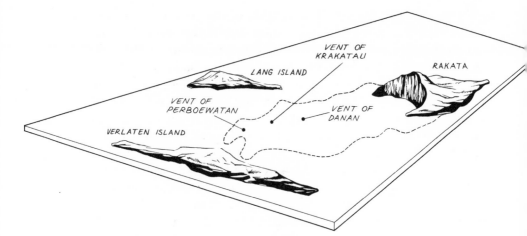

Fig. 3.8   Remainder of Krakatau

The Mystery of Atlantis

The legend of Atlantis, a mythical island kingdom described by the Greek philosopher Plato as a "utopia", has long been suspected of having been based on the history of the island of Santorini (or Thera), the Minoan settlement in the Aegean Sea. The Minoan peoples had inhabited Crete since the third millenium B.C., erecting there the great palaces of Knossos and Phaistos and developing linear writing from early hierogliphic writing. Yet for unknown reasons, the Minoan civilization rapidly faded into poverty and obscurity after about 1500 B.C.

Plato provided a clue to this mystery when he wrote: "there occured (in Atlantis) violent earthquakes and floods; and in a single day and night of misfortune . . . the island . . . disappeared in the depths of the sea". Today's visitor to the peaceful island of Thera finds another clue to the mystery of Atlantis's disappearance when noticing the conical hill rising from the moon-shaped harbor, a clear indication of a **caldera** from an ancient volcano. Such a large cone must have been the site of a catastrophic eruption spewing massive amounts of ash into the atmosphere.

Moreover, reported Plato, following the "earthquake" the sea around Atlantis "became an impassable barrier of mud to voyagers sailing from hence to any part of the ocean". Such a mud barrier could have been the consequence of a heavy fall of ash mixed with water and proof of the correctness of this assumption was uncovered late in the nineteenth century when volcanic ash—a key ingredient in the manufacture of the waterproof cement needed for the construction of the Suez Canal was discovered in the sediment of the eastern Mediterranean, in Egypt's Nile delta, in parts of the Black Sea *and* on Thera. Excavation of Thera's ash deposits, which were as deep as 270 m (900 ft.), led to the discovery of the buried city of Akrotiri and to the well-preserved wall paintings documenting in vivid detail the Minoan way of life.

There remained little doubt then that a destructive explosion had, some time in the past, destroyed Thera, but precise dating of the event remained elusive until 1994 when, deep in central Greenland, scientists drilled and recovered ice cores that are providing clues to historic events. Such cores, drilled as deep as 2 700 m (9,000 ft.), are silent witnesses to natural events that occurred in the Northern Hemisphere during the last nine thousand years, starting at a time when the great ice sheets were melting. The age of the ash layers found in the ice cores have been identified by their sulphur content and over 80 percent have been matched with known volcanic eruptions of the last two thousand years (such as that in A.D. 79 at Vesuvius). By using the known eruptions to establish a time scale, the destruction of Thera has been definitively dated at 1623 B.C., a solid link to the legend of Atlantis. Modern science and technology have thus provided a solution to the mystery of Atlantis and verified the "hard to believe" statements of Plato.

But what happened to the Minoans? No one can state their destiny with certainty, but it is reasonable to assume that, following the destruction of Thera, because of changes in weather, in the composition of the soil and in the riches of the sea, crop failures and a reduction of fish may have caused the beginning of a slow decline of the golden Minoan civilization.

Geothermal Energy

It may be fitting to end this rather depressing chapter with a look at a very old source of energy that is of growing importance to our future energy needs.

Volcanoes and earthquakes develop tremendous amounts of energy that, if it could be marshaled, would satisfy the world's increasing energy demands. Unfortunately, earthquakes are too random in location and too unpredictable in behavior to be tapped and volcanic eruptions too violent, but there is hope that volcanic heat may be utilized as a *geothermal energy* source.

Even our forefathers dreamed of harnessing volcanic heat, not during eruptions of course, but by capping boiling springs and *fumaroles,* the fuming vents often found in volcanic areas: they go deep enough into the earth to allow the underground water to reach the rocks heated by the underlying magma and become steam. The farmers of Ischia, the "smoking" island in the Gulf of Naples, used fumaroles to warm their potatoes in winter; all over the world, natural hot-steam baths have been used since antiquity for their curative effects and on Iceland, the highly volcanic island in the North Atlantic, most of the buildings are heated by underground steam.

The enormous potential of geothermal energy has been exploited for industrial purposes only in relatively recent times. In 1904, for the first time in the world, the fuming vents of Larderello in Tuscany began powering electric generators with the natural steam that, until then, had been only used for the extraction of boric acid. Originally, the electric energy thus obtained was sufficient to serve a limited local area, but at the present time, the Larderello and nearby Mount Amiata installations produce 1.7 percent of Italy's electric energy, replacing costly oil imports.

The second largest fields of geothermal energy in the world are now being exploited at Ohaaki and Waikirei, New Zealand, where two additional fields will soon join them. The Ohaaki installation is distinguished by its development of modern technology "in harmony with tribal values" of the native Maori. Water is piped through the underground steam, causing the water to boil into the steam used to run the turbines; as the steam gives up its heat to the turbines, it is converted back into water, starting the cycle over again.

The largest fields of geothermal energy in the United States started operating in 1960 in the area of "the Geysers" at Sonoma, California, potentially the largest source of geothermal energy discovered so far in the world.

As of this writing, geothermal energy furnishes the world with a negligible fraction of its total energy needs, but in view of the predicted early exhaustion of our oil reserves, it is getting serious consideration by all nations where it might be available. In 1970

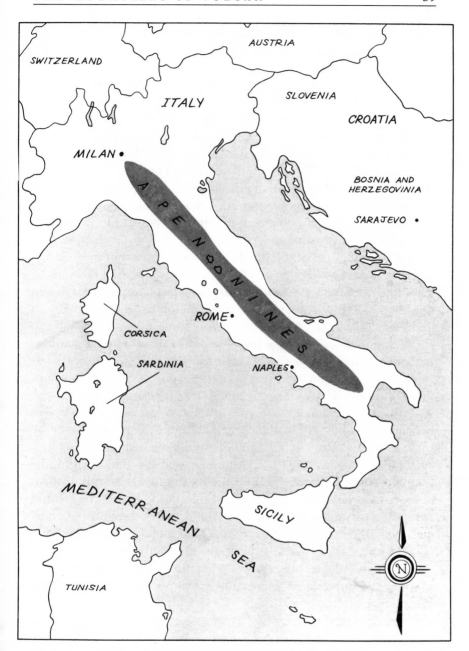

Fig. 3.9   Italian Geothermal Sources

the United States enacted the Geothermal Steam Act, which recognizes the development of geothermal energy "as a national goal". Italian governmental agencies have explored the entire area of central Italy west of the Apennine Mountains (Fig. 3.9) with wells as deep as 3 300 m (10,000 ft.) and discovered that the entire area is a promising source of the new energy.

Direct use of geothermal heat for heating homes and hot water has been developed in Japan, the territories of the former USSR, Bulgaria and Hungary. Small installations have also been built in Boise, Idaho, and Kalmath Falls, Oregon. According to the U.S. Geological Survey, geothermal energy at present supplies 1 percent of U.S. power and could supply up to 10 percent in the future. Geothermal energy already supplies 7 percent of the electricity in New Zealand, 21 percent in the Philippines, 18 percent in El Salvador and 11 percent in Kenya. Untapped resources are known to be available in Bolivia, Costa Rica, Ethiopia, India and Thailand, and may be available in two dozen other countries, including Brazil and Pakistan.

Making use of all the world's superficially available geothermal energy to meet our need for electricity would certainly give us time to develop nuclear fusion energy from hydrogen. Development of this unlimited form of energy is being carried out in many scientifically advanced countries of the world and is expected to become available within the early decades of the twenty-first century. But the supply of superficially available heat energy is limited and hence exhaustible. Although drilling deeper into the magma would certainly give us an inexhaustible supply of heat energy, magma's high temperature—2 800° C (5,000° F)—would melt the drills needed to reach it even if they were made of tungsten steel, the most temperature-resistant metal available today.

Yet our rapidly advancing materials technologies suggest to the optimists that we should not give up this wonderful dream of tapping the unbounded energy of the deep magma. "Hope is the last goddess to die", said the Romans and, we add, the best road to success.

4

Lisbon, 1755

> Why could it not have burst forth in the
> midst of an uninhabited desert? Why is
> Lisbon engulfed while Paris, no less
> wicked, dances?
>
> —Voltaire

Seventeenth-century Portugal owed its greatness to the sea.
Thanks to the adventurous travels along the Atlantic coast of
Africa sponsored by Prince Henry the Navigator, the rounding of
the Cape of Good Hope by Bartolomeu Dias in 1488, the opening
of the route to India by Vasco da Gama in 1497 and the "discovery"
of Brazil by Pedro Alvares Cabral in 1500, by the year of the Lord
1750 the small kingdom of Portugal had become an empire
extending from Africa to the Americas and Asia. Free of the Spanish yoke and of the Moorish invaders, Portugal was an absolutist
monarchy under the weak king Joseph I, ruled de facto by a pro-
gressive although brutal dictator, the Marquês de Pombal, who
had successfully fought the nobility, abolished the Cortes (the par-
liament) and opposed the influence of the Catholic Church, repre-
sented by the powerful Jesuits.

Lisbon, the capital of the kingdom and one of the most vital
commercial centers in Europe, was a delightful city of 275,000 peo-

ple, where the magnificent palaces of the nobility stood next to churches and monasteries, more numerous than in any other European city with the possible exception of Rome. The Alfama, the city's old quarter of Roman and Moorish origin, was terraced up the side of the eastern hills and was crowded with modest dwellings set along steep, winding alleys. A dominant feature of the city was—and still is—a rocky hill surrounded by the Castel de São Jorge, a Moorish citadel. The city, whose ancient name, Olisipo,[1] refers to a mythical city founded by Ulysses, was no stranger to earthquakes. Yet, although the Cathedral of Se Patriarcal, founded in 1150 by Alfonso I, had been wrecked in 1344 by an earthquake before being rebuilt in 1380, the residents of Lisbon were unprepared for the events that unfolded on All Saints' Day, November 1, 1755.

At about 9:30 A.M., at some distance west of the city, the strain accumulated over the centuries on one sector of the alpide belt (see p. 24) was suddenly released and, from a point below the earth's surface, called the *focus*, three great shock waves radiated out in rapid succession, demolishing all the houses in the lower city. The cathedral's dome was shattered and its roof and belfry subsequently burned, leaving only the choir and facade standing. A Gothic church, the revered fourteenth-century Igreja do Carmo, suffered horrible damage; only the apse, the pillared aisles and the outer walls were left standing. The entire city had become an instant ruin.

Within minutes, the shock was felt as far as Fez, Algiers, Madrid and Strasbourg (Fig. 4.1), while news of the disaster took almost two weeks to reach London. The shock was so powerful that the waters of Loch Lomond in Scotland rose and fell almost a meter (3 ft.) and those of the inland lakes high in the Alps oscillated, sloshing back and forth in what is called a **seiche.** As if Lisbon had not been punished enough, it would be struck two more devastating blows.

Lisbon lies on the right bank of the Tejo River estuary, near the area where the river empties into the Atlantic Ocean. The city stretches for many kilometers along a wide channel, the Entrada do Tejo, and a tidal lake, the Rada de Lisboa, near the lower reaches of the estuary (Fig. 4.2). As the shaking of the earthquake subsided, the sea receded, laying bare a sandbar at the mouth of

---

[1] An abbreviation of *Olisipo*, Greek for city of Ulysses.

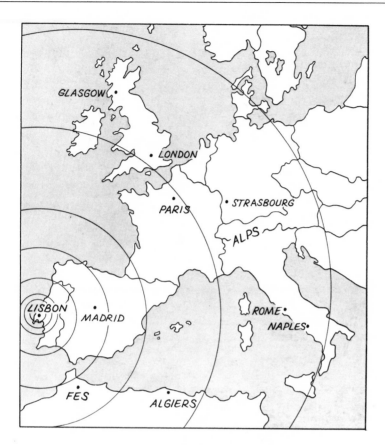

Fig. 4.1   Seismic Waves Radiate from Lisbon

the estuary. The sea returned as a **tsunami,** a wave 12 m (40 ft.) high, which ran up the estuary at incredible speed, broke over the quays and wrecked all the ships along the Tejo. The waters of the river then rushed toward the Lisbon harbor in three successive waves, throwing the anchored boats onto the land, crushing the docks and razing all the buildings on the city's main square, the Terreiro do Poço. (Because of its speed, when a tsunami strikes land, it does so as a crashing wave many times higher than its height at sea and causes damage far inland [Fig. 4.3]).

Great waves from the tsunami continued advancing and receding for two days, traveling the length and breadth of the Atlantic, with heights of as much as 3 m (10 ft.) as far north as Holland and 4 m (13 ft.) as far west as the Caribbean islands of Antigua and Martinique.

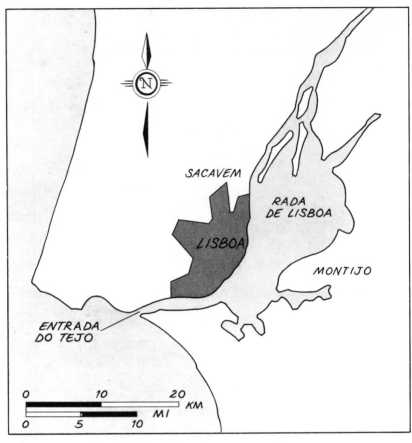

Fig. 4.2   The Environs of Lisbon

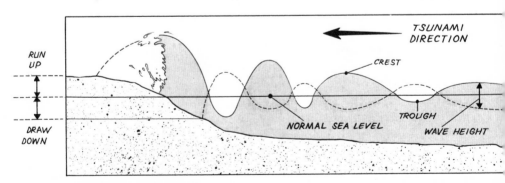

Fig. 4.3   Tsunami

Finally, at noon on November 1, as a cloud of dust covered the ruined city and changed the bright morning into a murky evening, a last shock was felt in the northern section of town and, shortly thereafter, fires raged everywhere. The Royal Palace, the recently completed Opera House and the magnificent cathedral (which, although damaged, had survived the earthquake), were consumed by fire.

After the fires had devoured what was left of the city, the fearful survivors of the Lisbon disaster gathered in squares and streets, eager to hear the prophecies of priests, nuns, friars and monks. The survivors were told the wrath of God was punishing them, the corrupt people of Lisbon, because they had given themselves to lives of sin, attending theaters and even bullfights, rather than listening to the admonitions of the church. Yes, they, and even some people of the cloth, had fornicated instead of praying in the churches, and this was why God had been compelled to destroy His own sacred houses. The time of penance was here: the continuing *aftershocks* were proof, if any was needed, that Almighty God was admonishing his flock to change behavior or expect the worst. Such was the fear of further punishment that, every omen, every one of the five hundred aftershocks felt during the following year, enhanced the impact of the clergy's menacing words and nurtured the fears of the nobility as well as of the general populace.

In the terrifying aftermath, one man used his power and influence to bring sanity back to the citizenry: the Marquês de Pombal wanted the buildings repaired and the dead buried before the plague struck. Unfazed by the disaster, he had a dream: to see the city rebuilt on a plan more magnificent than that of the destroyed capital. He was a brutal dictator but also a disciple of the French Enlightenment, an admirer of Voltaire and the Encyclopedists. He conceded that nothing could happen on this earth without the will of God, yet he believed that the earthquake had not been the consequence of divine punishment but a natural phenomenon, and that the present was a time to build rather than to listen to the prophets of doom. The door to seismology, the scientific study of earthquakes, had been opened by this hated but extraordinary man.

Within a few years, Lisbon was rebuilt, more elegant and superb than it had ever been, becoming to this day a tourist Mecca to visitors from all over the world.

5

The Birth
of Seismology

There stood the stout one-hoss-shay
As fresh as on Lisbon-earthquake day.

—OLIVER WENDELL HOLMES

Earthquakes have been variously described by Greek and Roman scientists and philosophers whose countries were often victims of these natural scourges. Pliny the Elder, writing in A.D. 200, believed that they were Mother Earth's way of protesting the violation of her domain by man's wickedness in mining for gold, silver and iron. Whether recorded or not, earthquakes have occurred since the beginning of time, but it was not until the middle of the eighteenth century that their study achieved scientific stature and moved out from under the cloak of mythology and superstition. The catastrophic earthquakes of 1755 in Lisbon and of 1783 in the Calabria region of southern Italy were among the earliest to be thoroughly documented by scientists who dared to propose rational explanations of these terrifying events, thus initiating the field of *seismology*, the science of earthquakes.

Waves

We are all familiar with the phenomenon of the disturbance generated by dropping a stone in the still waters of a pond: the energy of the dropped stone creates a circular, slowly expanding wave that eventually uses up its energy in friction (Fig. 5.1). What is typical of such a phenomenon is that the wave moves outward while the water particles of the pond move only up and down, and remain in the same location.

Various other natural phenomena excite the same behavior: a wave from a sound source radiates through the air in all directions and light waves similarly emanate from a light source. In all these phenomena the medium through which the waves travel does not move, but is temporarily disturbed by the energy of the passing wave.

The motion of all these waves is characterized by three quantities: the period T, the time between the beginning and the end of a complete wave oscillation; the amplitude A, or magnitude of the wave's oscillations; and the frequency f, or number of wave oscillations per unit of time (usually, per second) (Fig. 5.2).

While Fig. 5.2 accurately represents the behavior of waves, it cannot exhibit another of their characteristics: the direction of their motions. Some waves, like those in the pond, move the medium up and down at right angles to the direction of the wave motion, and are called *transverse waves;* other transverse waves oscillate horizontally. A second type of wave, such as a sound

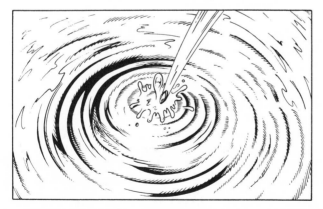

Fig. 5.1   Radiating Waves Caused by a Stone Thrown in a Pond

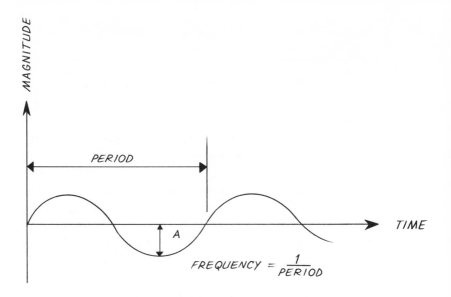

Fig. 5.2   Wave Diagram

wave, is called *longitudinal,* because it moves the medium back and forth in the direction of the wave motion, compressing and expanding the medium through which it travels.

A background on wave behavior is necessary to understand how the pent-up energy of an earthquake, which is released at a point *inside* the earth called the *focus* or *hypocenter,* reaches the earth's *surface,* often with disastrous consequences for us who live on it.

Seismic Waves

When two tectonic plates suddenly move with respect to each other, *seismic waves*[1] radiate through the earth from the focus in two types of so-called *body waves.* The P-waves, (for *primary*) or *pressure* waves, are the fastest. They are longitudinal waves, analogous to those generated by sound: they move through rock and soil at speeds of about 6 km / sec (3.8 mi. / sec.) and through water at about one-third that speed. Following the primary waves, the transverse *S* waves, the *secondary* or *shear waves,* travel *only* through solids at about 3 km / sec (1.9 mi. / sec.). Their lower speed

[1] From the Greek *seismos,* for earthquake.

is associated with a lower frequency of oscillation[2] and higher amplitudes of motion, making them more dangerous than the primary waves (Fig. 5.3).

As the body waves reach the earth's surface, they give rise to two types of *surface waves:* the longitudinal Rayleigh waves that travel leisurely on the surface of the earth with low frequencies and large amplitudes, undulating like long ocean waves, at speeds of less than 3 km / sec (1.9 mi. / sec.), and the Love waves, similar to the S waves in both character and speed, whose lateral and vertical shearing motion cause much of the ground shaking that is felt in earthquakes.

Moreover, when the body waves encounter the layers of different materials of which the earth is made (rocks, soils, liquids and gases), depending on the angle at which they hit the boundary between two different layers, they either bounce back, are *reflected,* as is our image in a mirror, or change direction, are *refracted,* as are light rays when crossing from air to water (Fig. 5.4). As earth-

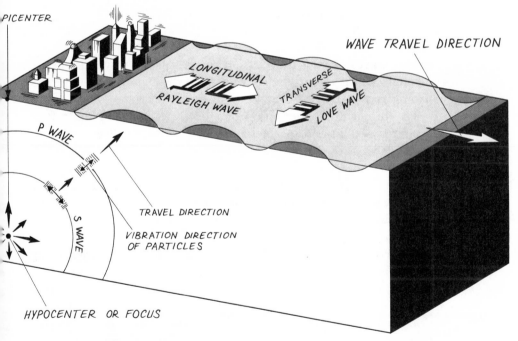

Fig. 5.3   Seismic Waves

[2] *Frequency* is the number of times per second that waves complete a full cycle before returning to their initial zero value.

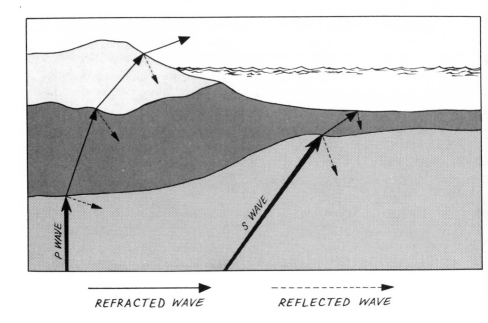

REFRACTED WAVE   REFLECTED WAVE

Fig. 5.4   Reflection and Refraction of Waves

quake waves travel they also change speed, traveling faster
through denser materials like rock than through looser materials
like most other soils.

A complicated interaction of waves results from the reflec-
tions, refractions and speed changes caused by the variable nature
of the earth's soils. To decipher this accumulated jumble of interacting waves, instruments called **seismographs,** which record the
relative strength or *intensity* and the duration of earthquakes, are
at present located throughout the world and record the arrival
time as well as the frequency and amplitude of each of the compo-
nent waves (separate instruments are required to measure vertical
and horizontal, N–S and E–W motions) (Fig. 5.5). By measuring
the difference in arrival time at a seismograph location of each
type of wave, the distance from the *epicenter* (the point on the
earth's surface directly above the focus) to the seismograph location can be approximately evaluated, and by establishing this distance from three observation locations, the epicenter can be
pinpointed (Fig. 5.6). Unfortunately, this method works only if the
distance from the epicenter to the seismograph station, when measured along the earth's surface, is less than about 11 000 km (7,000

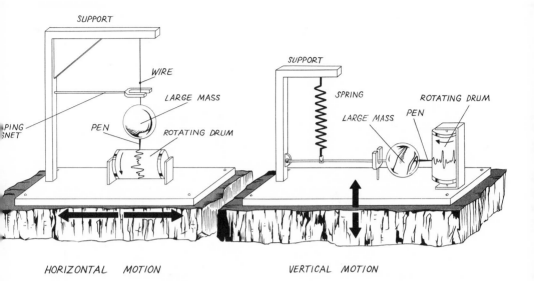

SUPPORT
WIRE
LARGE MASS
PEN
ROTATING DRUM
PING
NET

SUPPORT
SPRING
ROTATING DRUM
PEN
LARGE MASS
PEN

HORIZONTAL MOTION   VERTICAL MOTION

Fig. 5.5   Seismograph

mi.), because the S waves cannot cross the liquid part of the earth's core, while the other seismic waves traveling in a straight line can reach any distant point (Fig. 5.7).

The Aftermath of Lisbon

As many as fifty thousand people were killed by the crumbling structures, drowned by fierce tsunamis or burned by the citywide conflagration following the Lisbon earthquake. The destruction of the city was virtually complete yet the damage varied from section to section depending on the nature of the underlying soils. All the buildings constructed on soft, blue clays were totally destroyed; those sitting on harder sands and gravels suffered serious damage; but those sitting on rock, like limestones and basalt, were virtually undamaged. Similar findings were later reported by G. Vivenzio, a member of the Neapolitan Academy, after the 1783 Calabria earthquake. Vivenzio explained, for the first time, that in an earthquake the amount of structural damage depends essentially on the nature of the site: structures on the soft soils of the *alluvial plain* (consisting of slowly deposited river soils) are always more seriously affected than those on rocky hills. As evidenced by the 1989

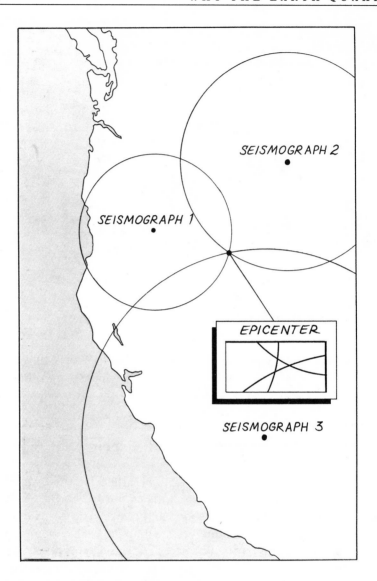

Fig. 5.6   Locating the Epicenter

Loma Prieta earthquake that devastated San Francisco's buildings constructed on soft soils or manmade ground, this lesson took over two hundred years to sink in.

Five years after the Lisbon earthquake, John Michell, a professor of geology at Cambridge University, suggested that the vibratory motions of earthquakes were due to the passage of waves

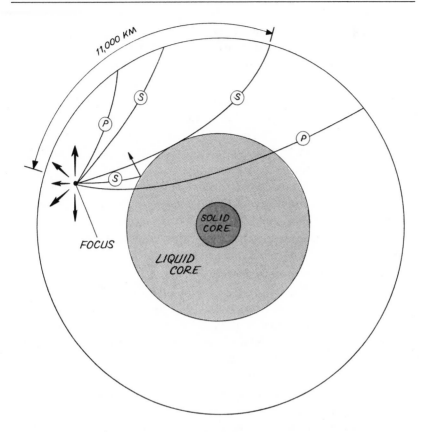

Fig. 5.7   Limit of Seismograph Measurements of Waves

through the earth's crust. He also observed that the location of the focus could be determined by measuring the direction and arrival time of the seismic waves from at least two observation posts and used this method to calculate the depth of the focus of the Lisbon earthquake from the epicenter on the earth's surface to be between 1.6 and 4.8 km (1 and 3 mi.). (We now know that the focus of earthquakes can be from a few meters to hundreds of kilometers below the earth's surface and that the deeper the focus, the less damaging the earthquake.)

The first catalogs of earthquakes were prepared early in the nineteenth century, and in 1843 the Frenchman Alexis Perrey began publishing annual earthquake lists that, by 1871, led him to overemphasize the linkage of the moon to the occurence of earthquakes. The subject is mysterious, although the influence of the moon on the earth is undeniable: the world's oceans rise and fall twice each lunar day and every time the moon passes over the island of Hawaii, the island rises by about 100 mm (4 in.) due to the gravitational pull the moon exerts on the earth. Whether the moon influences earthquakes remains a mystery, as does the fact that, for some unknown reason, earthquakes are more frequent in winter than in any other season, *except* for the most destructive earthquakes (see p. 49 for the lunar influence on volcanoes).

In 1846 Robert Mallet used the laws of wave motion in solids to explain earthquake phenomena (see p. 67) and in 1855 L. Palmieri constructed the first electromagnetic seismograph to measure the intensity of seismic events (he installed the instrument in the observatory at the base of Mount Vesuvius). In the 1880s, John Milne introduced accurate seismographs first in Japan, and then in England and all of its colonies. Thanks to this seismograph network, which has grown considerably since the 1880s, we can now explain how earthquake forces are transmitted throughout the earth and predict the location and strength of earthquakes, although not yet the time of their occurrence.

While the science of seismology was being developed, the population of the earth was rapidly increasing and the movement away from a rural economy brought more and more people to the cities. As a consequence of these factors, earthquakes that struck population centers became progressively more deadly and our ability to measure their intensity and to locate their foci as well as that of predicting their occurrence, became more and more urgent.

Manmade "Earthquakes"

Discovering the cause of a natural phenomenon does not necessarily banish a fear of it. That would require either ignoring it or curbing its power.

Perhaps we would be more inclined to put our shoulder to the job of taming earthquakes than to be lulled by a fatalistic acceptance of their danger, if we became aware that for almost half a century we have been able to excite our own, quite powerful earthquakes. Luckily, this well-kept secret has recently been revealed by a spectacular event in the People's Republic of China.

On December 28, 1992, at 12:50 A.M., Eastern standard time, Chinese army engineers blew up Potai mountain outside Zhuhai in Guangdong province (southern China), moving 4.9 million cubic meters (338 million cubic feet) of earth and stone in order to level the ground for the improvement of a local airport. The explosion required 11 000 tonnes (12,000 tons) of dynamite to flatten the mountain, and generated an earthquake of magnitude 2.5, according to the New China Agency. The thousands of technicians involved in this largest-ever dynamite explosion had spent months taking special measures to reduce the shock of the explosion from an expected magnitude of 4 to 2.5. But seismologists in Hong Kong, 100 km (62 mi.) from Zhuhai, measured the force of the seismic wave at magnitude 3.5, and the population of Macao, the Portuguese territory only 60 km (37 mi.) from the blast, strongly felt the explosion, although they did not report damage or injuries.

What makes this explosion significant is that it was as powerful as the first fission bomb dropped on Hiroshima in 1945, which had a yield of 10 700 tonnes (12,000 tons) TNT equivalent. This blast-generated "earthquake" is undoubtedly smaller than the undivulged "earthquake" caused by an underground Chinese nuclear explosion with a 605 000 tonnes (660,000 tons) TNT equivalent. Similar earthquakes have certainly been excited in the development of the many 40-megaton fusion bombs in the arsenals of the United States and the former USSR and now probably of other countries.

Many other manmade "earthquakes" have been observed. The city of Johannesburg, South Africa, is rocked periodically by tremors of measurable intensity. When first observed, these shocks were a mystery since there were no known faults in the area and no previously known seismic activity. Investigations carried out in the 1960s revealed that the city's foundation was laced with abandoned mines and that these would periodically collapse, setting off a small "earthquake."

The area around Denver, Colorado, also recently experienced a similar series of mysterious minor earthquakes. The cause was eventually discovered: in order to improve the yield of gas wells, water under pressure was pumped into the ground and may have acted as a lubricant, permitting layers of rock to slide, inducing a seismic shock.

Mankind can now boast of being able to match nature in creating earthquakes, but still remains virtually impotent, while relentlessly moving along the path of earthquake-hazard reduction.

6

Measuring Earthquakes

And behold, the curtain of the temple was torn in two from top to bottom; and the earth quaked, and the rocks were rent.

—Matthew 27:51

The first question we ask upon hearing of an earthquake is: "Where did it happen?", and, in almost the same breath: "How big was it?" But each of us interprets the *intensity* of an earthquake subjectively and what feels severe to one person may feel moderate to another. Being thrown out of bed in the middle of the night during an earthquake in Tokyo was a scary and unusual experience to one of us but, in the light of day, it turned out to be a frequent, moderate experience to the natives of the Japanese capital. Hence, it is obviously essential to measure earthquakes by objective, scientific means.

Qualitative Intensity

Before the availability of instruments capable of a quantitative measure of their magnitude, earthquakes were classified according to *intensity*, the qualitative reaction of people present at the event and to the observed or reported physical damage. In China such historic records go back to the time of the Ming Dynasty (1368–1644) and to a catalog dating from 780 B.C. In 1783 a scale describing in a more objective manner five earthquake intensities was first proposed in Italy by Dr. D. Pignataro, a physician. In 1883, M. S. Rossi and F. A. Forel proposed a ten-degree scale, the Rossi-Forel scale, still occasionally used in Europe, but the ten-degree scale proposed in 1902 by Giuseppe Mercalli was the first to find a wide consensus among seismologists because of its more detailed and explicit definitions. Finally, in 1931 seismologists around the world generally agreed to adopt as a standard scale the **Modified Mercalli scale,** which describes twelve degrees of intensity ranging from I (felt only by few people) to XII (total damage). Even with such scales, the subjectivity of the observer renders their use less than perfect, since even today experienced seismologists sometimes assign *different* intensities to the *same* earthquake.

Magnitude Scale

A quantitative measure of earthquake *magnitude* is given by the well known *Richter scale,* which is an index of the amount of energy released by the quake. Dr. Charles Richter of the California Institute of Technology proposed the first such scale in 1935, the so-called *local magnitude scale* (M_L), to standardize the measurement of earthquakes in southern California as recorded on a particular instrument, the Wood-Anderson seismograph. He suggested an evaluation based on two numbers: the first is the maximum excursion of the needle on the Wood-Anderson seismograph, which measures the *amplitude* of the maximum earthquake motion; the second is the square root of the ratio of the actual distance between the seismograph and the epicenter to a standard epicentral distance of 100 km (62 mi.). These two numbers are then related by a mathematical formula (using common logarithms of base 10)[1] to an index, M_L, now called the *Richter magnitude.* Since the formula is logarithmic, a *unit* change in the Richter scale represents a *tenfold* increase in the vibratory amplitude of the earth.

[1] A logarithm to the base 10 is the power x to which the number 10 must be raised to produce a given number, N. Therefore $10^x = N$ or $\log_{10} N = X$.

It is important to notice that an increase of one in the Richter magnitude M_L, does *not* really represent the corresponding tremendous increase in the energy of an earthquake: a more complex relation between energy release and magnitude proves that an increase of one unit in the earthquake's *magnitude* corresponds approximately to an increase of thirty-two times in the earthquake's *energy* E.

With his colleague Dr. Beno Gutenberg, Richter recognized the limitations of his first scale, which was designed to measure quakes only in southern California. They were thus led to develop other scales: the *body-wave scale*, m_b, to deal with deep-focus earthquakes, and the *surface-wave scale*, M_s, better suited to measure more distant and larger quakes.

With all these different scales, it is not surprising that the Richter magnitudes reported in the news media are confusing: they sometimes refer to the body-wave scale (for small to moderate quakes) and sometimes to the surface-wave scale (for large quakes). To add to the confusion, another scale—the so-called *moment magnitude scale*, M_w—was recently adopted to measure the world's strongest earthquakes. The moment magnitude scale takes into account both the energy release and the size (amplitude) of a distant earthquake. Based on this scale, the great Alaska earthquake of 1964, for example, is rated at over 9 whereas it was originally classified as 8.6 on the M_s scale. To simplify communication with our readers, we have chosen a single summary magnitude scale M, which we will simply call *the* Richter scale, defined as the magnitude that best describes a particular event, whether it is M_L for a local California earthquake, M_w for a large event or M_s for a distant event.

Magnitude and intensity scales measure different aspects of the strength of an earthquake. A high-magnitude earthquake may occur, as many do, below the ocean floor or in a sparsely inhabited region. Damage to structures near the epicenter of such a quake is obviously minor. But a low-magnitude quake with an epicenter in or near a large city can cause substantial damage and injury or death to city dwellers, and is better defined by the intensity scale. The 1960 earthquake in Agadir, Morocco, for instance, with a magnitude of only 5.75, destroyed the entire city and killed thousands of its inhabitants. Like most earthquakes causing heavy damage, its epicenter was within close range of a densely populated area, in this case, within 8 km (5 mi.) of the crowded alleyways of the

Kasbah; to further aggravate the event, the quake had a shallow focus (under 3 km [2 mi.]). The *isoseismal*[2] map of Agadir (such maps define regions of equal intensity), shows the extent of the wide-ranging destruction caused by the quake (Fig. 6.1). As we have seen, for a given magnitude, the intensity also depends essentially on the type of soil at the site: the intensity felt on soft ground is two to three times greater than that felt on rock, as we should have learned a long time ago from the Lisbon earthquake.

To put earthquakes in perspective for our readers, we can state

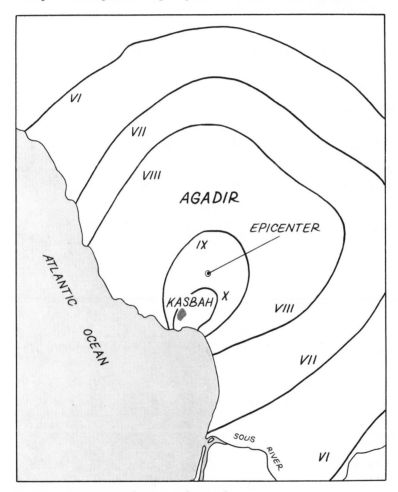

Fig. 6.1   Isoseismal Map of Agadir

[2] From the Greek *iso*, for equal.

that, throughout the world, a magnitude 6 quake is a daily occurrence, a 7 is a weekly occurrence and an 8 is an annual occurrence,
but that most of the high-magnitude shocks occur in sparsely populated parts of the world and therefore receive little attention. As
shown in Table 6.1, the strongest earthquakes are not necessarily
the most destructive and deadly; but often when a quake strikes a
city, the fire that follows causes the greatest damage. Such was the
case in Japan in 1923.

TABLE 6.1

Major Earthquakes of the Twentieth Century

Place	Year	Magnitude	Deaths
Afghanistan	1956	7.7	2,000
Alaska	1964	8.4	131
Algeria	1980	7.3	4,500
Armenia	1988	6.9	25,000
Bosnia	1963	6.0	1,100
California	1989	7.1	62
	1971	6.4	58
	1906	8.3	452
Chile	1960	8.3	5,000
	1939	8.3	28,000
	1906	8.6	20,000
China	1976	7.8	240,000
	1932	7.6	70,000
	1927	8.3	200,000
	1920	8.6	100,000
Ecuador	1949	6.8	6,000
Guatemala	1976	7.5	22,800
India	1950	8.7	1,500
	1935	7.5	30,000
	1934	8.4	11,000

Place	Year	Magnitude	Deaths
Iran	1978	7.7	25,000
	1972	6.9	5,000
	1968	7.4	12,000
	1962	7.1	12,200
	1957	7.4	2,500
Italy	1980	7.2	4,800
	1976	6.5	946
	1915	7.5	30,000
	1908	7.5	83,000
Japan	1948	7.3	5,100
	1946	8.4	2,000
	1933	8.9	3,000
	1923	8.3	100,000
Morocco	1960	5.8	12,000
Nicaragua	1972	6.2	5,000
Pakistan	1974	6.3	5,200
Peru	1970	7.7	66,800
Philippines	1990	7.7	1,700
	1976	7.8	8,000
Turkey	1983	7.1	1,300
	1976	7.9	4,000
	1975	6.8	2,300
	1970	7.4	1,100
	1966	6.9	2,500
	1953	7.2	1,200
	1939	7.9	30,000
Yemen	1982	6.0	2,800

7

The Kanto
Earthquake

> . . . you will be visited . . . with earthquake
> and great noise, with whirlwind and tem-
> pest, and the flame of devouring fire.
>
> —ISAIAH 29:6

For over twenty years early in the twentieth century, A. Inamura, a young seismologist at the University of Tokyo, had studied the pattern of seismicity along the Japanese islands and had predicted that there would soon be a very strong earthquake in the Kanto (eastern) district around Tokyo (Fig. 7.1). He publicized his prediction to both technical and governmental groups but was thwarted by the elder statesman among the world's seismologists, Dr. F. Omori, who thought Inamura's arguments were irresponsible. As a result, no action was taken by the Tokyo municipality despite Inamura's added forecast of the possibility of a consuming fire following the strong earthquake and of over 100,000 casualties.

Seismographs around the world sprang to life on Saturday, September 1, 1923: Kealakekua, Hawaii, recorded a "distant earthquake" at 7 P.M.; Berkeley, California, recorded a "very severe" earthquake 8 640 km (5,400 mi.) west of the station, with a duration of shocks and immediate aftershocks lasting almost four

Fig. 7.1   The Kanto District

hours; at West Bromwich, England, Mr. J. J. Shaw was awakened
at 4:11 A.M., when an alarm rang to signal the beginning of the
needle movements on the local observatory's seismograph; at the
Georgetown University seismographic station in Washington,
D.C., Father Torndorf, the director, noted the arrival time of a
great earthquake at 10:12 P.M.; it reached its maximum intensity
forty-five minutes later and lasted until three the next morning.
Dr. Omori, attending a scientific meeting in Australia, was observ-
ing a seismograph when he saw the trace of a major event and,
from the estimated distance, guessed that Inamura's prediction
had proven correct. From these observations the seismologists
decided that an earthquake of major proportions, with an epicenter located between Tokyo and Osaka, had taken place at noon
local time. Yet, apart from the traces on seismographs, the world

received no news of this event. How could such a seemingly catastrophic occurrence go unreported, given the availability of rapid
telegraphic and radio communications?

The Japanese islands are located along the western edge of the
Pacific plate, a region favored with both earthquakes and volcanoes (see p. 24). (The snow-capped peak of Mount Fuji, 3 776 m
(12,380 ft.) high and 96 km (60 mi.) southwest of Tokyo, is the best
known of a chain of two hundred volcanoes—of which fifty are
active—stretching from the South Seas to the Bering Sea. Since
the great eruption of 1707 blew up the Hoei crater of Mount Fuji,
the volcano had been quiet, although in 1923 steam still vented
from a point near the summit.)

Tokyo (known as Yedo before 1868),[1] on the east coast of the
main island of Honshu, is protected from the sea by Sagami Bay
at the head of the entrance to Tokyo Bay (see Fig. 7.4). To this day,
Tokyo is subject to weekly tremors and, until the early twentieth
century, some of its inhabitants believed that a giant fish under
the city would someday bring about its destruction. The city lies
on both sides of the Sumida River, with the oldest sections on the
western bank (Fig. 7.2). On a 40 m (130 ft.) high hill in the center of
the old town (Kojimachi-ku), sits the nineteenth-century Japanese-
style Imperial Palace surrounded by a moat. It occupies the site of
the residence of the former shoguns, which was destroyed by fire
in 1873.[2] A castle was first built on the site of the palace in 1457
but the importance of the city of Yedo dates from 1590, when the
Tokugawa family began a three-hundred-year rule.

Following World War I, the country experienced a period of
westernization and rapid industrialization. On the streets of
Tokyo, the jinriksha, drawn by one or two men, was being replaced
by trams, and in the Kojimachi-ku section near the palace, the residences of the feudal lords (daimyos) were replaced by government
offices. Six bridges were built to span the Sumida River,
expanding the boundaries of the city to 256 km² (100 sq. mi.) and
the population to five million (it is more than four times as populous today).

Facing Sagami Bay in Kamakura is the Daibutsu, the great

---

[1] The eastern capital of Japan as distinguished from Kyoto, its western
capital.

[2] From the Chinese, *chiang chun*, the leader of the army. Shoguns were dynas-
tic military governors of Japan who held greater political power than the emperor,
who was relegated to a ceremonial role.

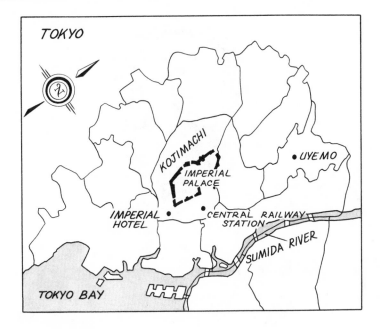

Fig. 7.2   Tokyo

fourteenth-century bronze Buddha, who sits in perfect serenity
with eyes "whose gaze penetrates into the very soul of each wor-
shipper standing before him" (Fig. 7.3).

On the morning of September 1, 1923, the volcano on Ō-shima,
an island 30 km (19 mi.) across Sagami bay, began belching smoke,
a warning to all who saw it that the earth was about to burst.
The floor of the bay started groaning and suddenly tilted. Then the
whole bay twisted in a clockwise direction, its north shore moving
about 3 m (10 ft.) southeastward and Ō-shima moving northeast-
ward about 3.6 m (12 ft.) (Fig. 7.4). This incredible wrenching of
the earth's crust sent a shock wave from a focus 48 km (30 mi.)
below the bay, a wave that radiated outward and destroyed nearly
everything in its path.

Only the great Buddha, Daibutsu, serenely rode out the storm.

Soundings taken of the seabed at a later date showed that the
area near the north end of the bay had risen an incredible 450 m
(1,500 ft.) and that barely 2 km (1 mi.) further south, the ground
had dropped 1 440 m (2,400 ft.). As a consequence of these major
seabed changes, a new island 48 km (30 mi.) long appeared off the
coast of Yokohama and one of Japan's jewels, the island of Enos-

Fig. 7.3   Daibutsu of Kamakura

hima, disappeared in the sea. In Yokohama, on the western shore
of Tokyo Bay, water spouts burst from low-lying areas, turning the
earth to mud, a common occurrence when quakes occur in areas
with water near the earth's surface, where the shaking of the crust
may cause **liquefaction** of the soil, ejecting water, sand or mud and
building cones, similar to those resulting from volcanoes (see p.
36), with diameters of up to 30 m (100 ft.).

Only seconds before the shock wave struck Tokyo itself, the
great city of Yokohama was shaken to the ground. "It came in a
dozen terrible vertical movements, the earth rising and falling
beneath us", reported one witness. Roderick Matheson, the corre-
spondent for the *Chicago Tribune*, reported: "Three minutes before

noon came a grinding blow beneath our feet, the earth groaned, buildings began to shift and crack, and then the first of a series of tremendous shocks came with a roar. The ground swayed and swung, making a foothold almost impossible, while a fine dust rose from every building, darkening the air. The groaning of the swaying buildings rose to a roar and then to a deafening sound, as the pitching, swaying structures began to crumble and then fell". When the quake struck, the British attaché, Major R. E. Smith, and his wife were on the third floor of the embassy, sitting at a desk with hands clasped across the table. The building shook and began to crumble, carrying the couple down two stories with their hands still clasped. Miraculously, they landed uninjured.

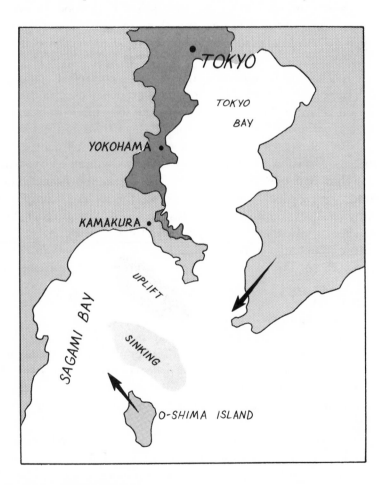

Fig. 7.4   Sagami Bay

The quake, of Richter magnitude 8.3, devastated a region of 115 000 km² (45,000 sq. mi.), damaging 70 percent of the cities, towns and villages within a 200 km (125 mi.) radius from Yoko-hama. Striking down royalty and commoner alike, almost 100,000 people perished and another 150,000 were injured or reported missing. All communications from the affected area were instantly severed, explaining the delay with which news of the disaster reached the outside world. Many apparently substantial structures shed their brick skins, which rained down on the fleeing citizens. Multistory buildings, with structures as flimsy as stage sets, as well as small, typical Japanese houses with heavy tile roofs, col-lapsed on themselves, trapping their occupants. Modern steel-framed and reinforced concrete 'skyscrapers' withstood the quake, although some showed fissures on their masonry facades, espe-cially at the third floor (see p. 116 for an explanation of this curious phenomenon).

As houses throughout the city of Tokyo collapsed, gas pipes broke, starting fires fueled by the wood and paper used in the con-struction of most Japanese dwellings at the time. (It has been reported that because of the time of the shock, two minutes before noon, the many cooking fires used to prepare lunch contributed to the conflagration.) The flames spread rapidly, fanned by strong winds that had been blowing since the early morning. Under-ground water pipes were severed by the quake, completely destroying the water distribution system and leaving firemen help-less to control the many fires that swept through the city.

More than half a million houses were destroyed and an area of 64 km² (25 sq. mi.) was burned to the ground. Within the first month following the great quake, 1,256 aftershocks were recorded. The first of these was severe enough to cause further extensive damage, particularly to buildings already weakened by the initial shock. But the ancient wooden pagoda in Ueno Park, north of the city, survived both the shock and subsequent fire, while more mod-ern buildings collapsed, because a properly braced and carefully connected wooden structure usually stretches and sways in response to an earthquake, behaving essentially in an elastic manner.

If this were not punishment enough for the tattered citizenry, the quake spawned a tsunami that roared into Tokyo Bay, swamp-ing the port of Yokohama and leaving it a mass of tangled wreck-age. The tsunami then snaked its way up the Sumida River, drowning many and washing away shoreline houses, before receding and beginning a long journey to distant shores around the Pacific Ocean. On September 4, its 6 m (20 ft.) high waves slammed onto the southern California coast, taller than any previously experienced by local mariners.

On the day following the quake, with aftershocks continuing to shake the region, fires continued to burn out of control, until, luckily, on the third day, a soaking rain helped to extinguish the flames, which were lapping at the little that was left to burn. In the aftermath of the conflagration, the prince regent (who three years later was to become Emperor Hirohito), returned to the damaged but intact Imperial Palace and, in the wake of the food shortages and the riots that broke out in various sections of the devastated cities of Tokyo and Yokohama, tried by his presence to calm the terrorized citizens. He led the relief effort and even ordered the palace gates to be opened to refugees. In the near-panic conditions following the quake and fire, some residents blamed expatriate Korean workers for setting the fires. Many Koreans were arrested and interned, although they were released after calm and order were restored.

The Aftermath

After the quake, one of the first official edicts mandated by the city administration was to rebuild Tokyo with wider streets, not just to accommodate the growing traffic but to provide a firebreak for the next big earthquake that would strike. For years following the quake, another safety measure limited the height of buildings in the Kanto district to 31 m (103 ft.), although today Tokyo has over 150 skyscrapers exceeding 100 m (333 ft.) in height.

8

Resisting Earthquakes

If an earthquake were to engulf England tomorrow, the English would manage to meet and dine somewhere among the rubbish, just to celebrate the event.

—WILLIAM BLANCHARD JERROLD

Learning from the Earthquakes

When three violent earthquakes with estimated magnitudes of 8.6, 8.4 and 8.7 struck New Madrid, Missouri, in the central United States in the winter of 1811, no structure larger than a log cabin stood in that sparsely settled region of the newly annexed Louisiana territory. Of course, chimneys were knocked down and even the somewhat limber wooden cabins were so badly damaged that they were abandoned by their pioneer occupants. The earthquakes were so powerful as to be felt as far away as Boston, 1 760 km (1,100 mi.) to the east, and to cause bricks to fall off structures in Georgia and South Carolina (Fig. 8.1). These little remembered quakes, apparently caused by an adjustment to a fracture in deep-seated rock in the center of the North American plate, resulted in negligible loss of life, both because the region was so

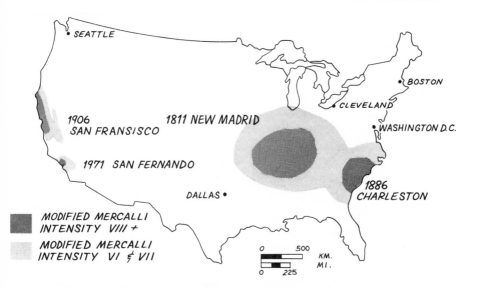

Fig. 8.1   Intensity of Major U.S. Earthquakes

sparsely inhabited and because the settlers' log cabins were particularly well suited to resist the shocks by their elasticity.

Every historic earthquake provides us with a fresh view of
ground shaking and leads to new knowledge of the resistance that
needs to be built into our structures. A committee, appointed by
the Neapolitan Academy after the 1783 earthquake in southern
Calabria, first observed the different extent of damage to a structure depending on its siting on solid rock or soft ground (see p.
79). One hundred years later, another scientific committee (whose
report was suppressed by the authorities for fear it would damage
the stature of the young and growing city) studied the 1868 earthquake in San Francisco and learned that the greatest damage had
occurred in that city's business district, the Yerba Buena cove,
which had been filled with dump material, and where the ground
had sunk as much as 600 mm (2 ft.) due to the consolidation from
the shaking of the loosely packed material. The imperial committee investigating the 1891 quakes at Mino and Owari, Japan, studied the methods of predicting earthquakes and proposed designs
for earthquake-proof buildings. One year after another large and
destructive earthquake in San Francisco, that of 1906, the Caribbean island of Jamaica had its own event, one that claimed over a
thousand lives, and led to the enactment of one of the world's first

building laws defining antiseismic construction. (Kingston, Jamaica, had previously suffered a major earthquake in 1692 when half the capital, home to pirates and known as Port Royal, submerged 10 m [30 ft.].)

Ground and Building Motion

Most loads a building must support, such as the weight of floors, walls and occupants, are applied slowly over time and are called *static* loads. But the load or force of an earthquake is applied and changes in magnitude very rapidly: it is called a *dynamic* force.

The back-and-forth or up-and-down oscillations of the ground near the epicenter of a quake are generally very violent and quick. The time it takes to complete one cycle of such a motion is called the *period* of the oscillation and is particularly short near the epicenter, usually less than one second. As the seismic wave travels away from the focus, the earth acts somewhat like a shock absorber and *damps* out the short-period vibrations, leaving intact the waves dominated by long-period oscillations. Both near and far from the epicenter, the prevailing vibrations are those with the largest *amplitude*, the largest displacement of the back-and-forth or up-and-down motion (Fig. 8.2). Amplitudes of earthquakes may vary from 1 mm (0.039 in.) in a mild shock, to 70 mm (2.8 in.) in a somewhat destructive earthquake and 220 mm (8.7 in.) in a devastating quake.

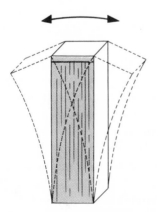
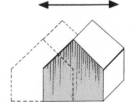

Fig. 8.2   How Buildings Shake

A structure on or in the earth vibrates in response to the seismically induced motion of the ground, as a string vibrates when
plucked by a finger. And, just as differently tensioned strings emit
different sounds when plucked, structures with physically different properties respond differently to the shaking of the earth: a
rigid structure will tend to move together with the movement of
the ground while a flexible structure will lag behind it (Fig. 8.3).
The property that characterizes rigid or flexible buildings is their
natural period of vibration, that is, the period of the *free oscillations*
of a building after the quake movements are over (approximately
0.1 seconds per story). Low buildings, especially if built with
masonry walls, tend to be stiff and to have short natural periods,
while tall buildings (with steel or concrete frames) tend to be flexible and have long natural periods. Along a typical city street,
buildings of different heights and construction are lined up next
to each other. Responding to an earthquake, these buildings will
vibrate with different periods and different amplitudes, and if
built too close together, may even bump into each other (Fig. 8.4),
causing their facades to crumble. A modern city like Tokyo, in a
seismically active region, requires buildings to be separated from
each other by a space sufficiently large that each can oscillate
freely without bumping into its neighbor.

Fans watching a World Series baseball game at San Fran-

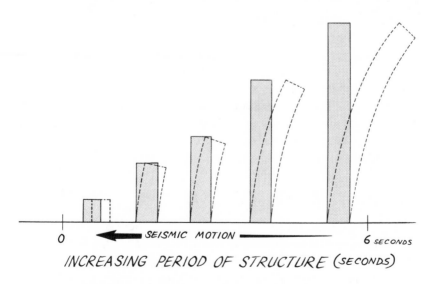

INCREASING PERIOD OF STRUCTURE (seconds)

Fig. 8.3   Period of Vibration of Structures

Fig. 8.4   Two Buildings with Different Periods Bump Together

cisco's Candlestick Park were unwilling witnesses to the 1989 Loma Prieta earthquake. The flexible structure of the stands shook briefly as the seismic waves traveled from the underlying rock through the foundation pilings to the structure. Much to the relief of the fans, after a few oscillations, the structure returned to its original position without causing significant damage.

Since buildings must be designed to anticipate earthquake shocks coming from any direction the plan of a structure affects its response. A structure with a "regular" plan (square, circular, triangular or otherwise symmetrical), will vibrate without *twisting*, while one with an irregular plan will both twist and move laterally when "plucked" by the earthquake (Fig. 8.5). A tendency to twist generates a complex response of the structure and requires particular structural characteristics to successfully resist a quake.

When the natural period of vibration of a building happens to be equal to the ground's earthquake period for several cycles, as sometimes happens, the motions of the building increase with each vibration cycle and the building is said to be in *resonance* with the ground. This phenomenon is particularly dangerous for tall or flexible buildings where, if in resonance, the amplitude of vibrations can keep increasing, as a child's swing moves higher and higher with each "well-timed" (in resonance) push. Resonance was the essential cause of the collapse of 50 of the 124 spans of the Cypress Viaduct in Oakland, California, during the 1989 Loma Prieta earthquake (Fig. 8.6). The reinforced concrete frames of those particular spans, founded on weak soils, were in vertical res-

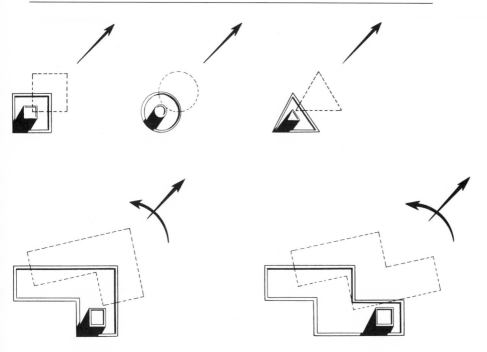

Fig. 8.5 How Shape Affects Buildings' Movements in an Earthquake

Fig. 8.6 Collapse of the Cypress Viaduct

onance with the ground's earthquake movements. As a consequence, the spans suffered increasing vertical motions and were in
effect slammed down, dropping the upper roadway on the lower
one and killing forty-two occupants trapped in their vehicles (two-
thirds of all the victims of the Loma Prieta quake). The remaining
spans, sitting on firm soil, remained standing.

The most important structural characteristic in preventing
earthquake damage is that of *ductility*, the property of a material
to bend, stretch and twist without breaking. For example, a ductile
structure of steel, under the large oscillating forces of a major
earthquake, is stressed *beyond* its limit of elasticity, in alternating
swings from tension to compression, but does not collapse. These
cycles of *plastic* (beyond the elastic) stresses, also known as *hysteretic cycles* (Fig. 8.7), absorb large amounts of energy that prevent
collapse and result in permanent, but usually acceptable, deforma-

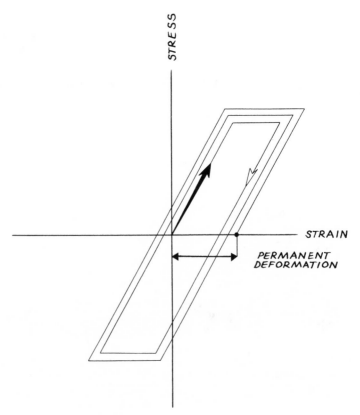

Fig. 8.7   Hysteretic Cycles

tions. In this context, steel, with its high ductility, is an ideal material in high-damage earthquake zones, while reinforced concrete, an otherwise remarkably inexpensive structural material, cannot meet the ductility requirements of the codes without substantial steel reinforcement. The columns of the Cypress Viaduct failed because their vertical steel reinforcing bars were not adequately tied together and bulged outward under the vertical seismic forces, eliminating the columns' ductility (see Fig. 8.6).

The Imperial Hotel

When Japan was reaching out to the world after the First World War, the great American architect, Frank Lloyd Wright (1867–1959), was asked to design a hotel to be financed by Baron Okura and the royal family. The Imperial Hotel was completed in 1921 in a style sympathetic to Japanese architecture, using handmade bricks and *oya*, a porous volcanic limestone (Fig. 8.8). In keeping with Wright's principle of resisting earthquakes by means of flexibility rather than by rigidity, the building's structure stood on two thousand concrete piles driven deep into the earth. Wright conceived the structure as a "super-dreadnought, floating on the mud as a battleship floats on salt water". The building was considered so unusual that it was not initially approved for construction by the doubting Japanese building authorities, but Wright, with his usual modesty, convinced them that "a world-famous architect would never come to Japan to erect a building that might fall down."

Newspapers prematurely announced that Frank Lloyd Wright's Imperial Hotel had been destroyed by the 1923 quake. One of the two hundred guests who was attending the hotel's official opening when the quake hit said: "It shook like a leaf. Some of us plunged through broken windows, but the building remained standing." (Wright was vindicated when the building survived without serious damage, but to his chagrin, this "miracle" went unrecognized in post-earthquake damage reports.) Arato Endo, Wright's Japanese colleague, wrote to him after the earthquake in 1923: "Now your chance is here. You will be received here now with admiration and appreciation, late, yes, but not too late. The whole city is at your disposal. You will have more appreciation here than in America." Wright, who had been dismissed in 1922 for

Fig. 8.8   The Imperial Hotel

his failure to prevent delays and cost overruns in the construction of the hotel, never returned to Japan. His Imperial Hotel succumbed in 1968, not to an earthquake but to the wrecker's ball, to make way for a large nondescript skyscraper on land that had become too valuable to accommodate such a modest, although superbly elegant, structure.

9

The Great Fault

I felt no trace whatever of fear; it was pure delight and welcome. "Go it," I almost cried aloud, "and go it stronger!"

—WILLIAM JAMES

The San Andreas Fault

The coast of the western United States, and particularly its sections in California and western Nevada, spawn 90 percent of the seismic activity in the United States. Moreover, this activity is more frequent and violent than elsewhere in the country because it is generated by shallow-focus earthquakes (less than 24 km [15 mi.] below the surface) and occurs along known rupture zones called *faults*. The region is sliced by the *San Andreas Fault* (Fig. 9.1), an intriguing 1 300 km (800 mi.) long fissure that starts at the Salton Sea in southern California, continues northward, caresses Los Angeles, crosses the city of San Francisco and dives briefly into the Pacific Ocean at Point Arena (175 km [109 mi.] northwest of the city) before reappearing at Punta Gorda in northwestern California, where it veers westward into the Pacific Ocean to join the Mendocino Fracture Zone. The fault—perhaps as much as twenty-

Fig. 9.1   The San Andreas Fault

nine million years old—is part of a system of fractures of the earth's crust along the boundary between the Pacific and the North American tectonic plates. The rock formations on both sides of the fault show a relative slipping displacement of as much as 560 km (350 mi.) in the last ten million years, due to the northwesterly movement of the Pacific plate relative to the North American plate (as if Los Angeles had moved up to San Francisco). Standing on the land on either side of the fault and looking at the land across it, the land on the other side appears to have moved to the right, which is why geologists call the displacement a *right-lateral strike slip* (Fig. 9.2).

The slip displacement along the fault takes place at a rate of about 50 mm (2 in.) per year. But, this movement does not take place uniformly along the length of the fault nor continually in time. One section of the fault may suddenly slip, causing an earthquake, while the rest of the fault remains locked until enough strain builds up and another section suddenly slips. Some fault sections, for example that from Salinas to San Luis Obispo (see Fig. 9.1), creep continuously, never storing up large strains and,

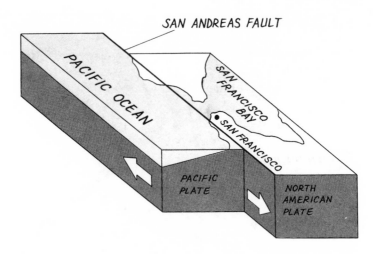

Fig. 9.2   Right Lateral Strike-Slip Motion

consequently, are only moderately active from a seismic view-point.

The San Andreas Fault, first identified by the geologist Andrew Lawson in 1893, is responsible for the most destructive earthquake witnessed by settlers after their arrival in the region. It occurred on January 9, 1857, at exactly 8:13 A.M., as evidenced by a clock that stopped in a San Francisco jeweller's shop. At that very same time, Professor George Davidson, lying in his north–south-oriented bed, was shaken slightly while his companion, in a bed oriented east–west, was thrown out of bed, proving that the principal earthquake shock had been north–south directed (Fig. 9.3). The earthquake was centered at Fort Tejon, an army post between Los Angeles and San Francisco, 6.4 km (4 mi.) from the San Andreas Fault, and caused a 64 km (40 mi.) long and 6 m (20 ft.) wide crack to open in the earth. The crack immediately clamped shut, leaving a *scarp*, the wound on the earth's surface. At one point along the scarp, a streambed was observed to have been displaced 8.7 m (29 ft.).

In 1868, an earthquake, called until 1906 the great San Francisco earthquake, resulted from movements along the Hayward Fault east of San Francisco Bay, one of California's many faults parallel to the San Andreas (see Fig. 9.1). In 1872 a shock along the San Andreas Fault in the region near Lone Pine, deep in the Sierra Nevada mountains, killed 10 percent of the town's population,

Fig. 9.3   Shaken Out of Bed on January 9, 1857

destroyed 90 percent of its adobe houses, and left a 9 m (30 ft.) deep depression between two cracks in the ground 75 m (250 ft.) apart. This last quake was the most powerful in the region until 5:12 A.M. on April 18, 1906.

San Francisco, 1906

At the turn of the twentieth century, the city of San Francisco was a rapidly growing metropolis. But, it was still primarily a city of modest dwellings, some separated and some sharing common masonry walls, and all with wood floors and roofs. More elegant mansions lined the streets of Nob Hill and imposing masonry-clad structures were centered in the business district on both sides of Market Street (Fig. 9.4). Where once upon a time only church spires punctuated the skyline, new buildings were rising everywhere to the unheard of height of twenty stories. *Semirigid* beam-to-column connections provided the necessary ductility for their steel skeletons to bend without breaking (Fig. 9.5). They were to be severely but successfully tested in the first decade of the new century.

The quake struck without warning at 5:12 A.M, unless we take as omens that three hours earlier, a journalist walking home and passing a livery stable, heard horses neighing and crashing their hoofs against the walls of the stalls; that dogs were heard barking

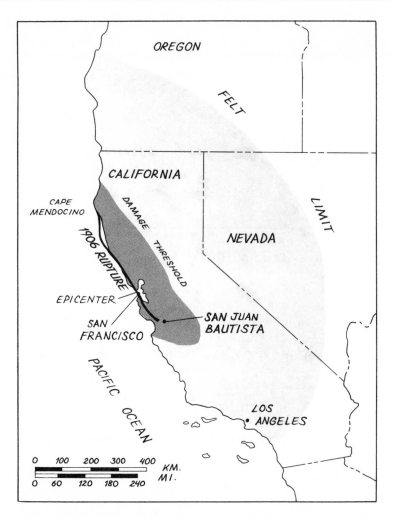

Fig. 9.4   Extent of Fault Rupture from the 1906 San Francisco Earthquake

throughout the city; and that twenty minutes before the quake, a horse pulling a milk cart was so excited it could not be calmed (see p. 126). The *precursors*, foreshocks with vibrations imperceptible to humans, that announce the clearing along a fault of obstacles preventing slippage and the release of major energy, sped across the earth at 10 km/sec (6.2 mi./sec.). Their signal reached the Tokyo seismographs only twenty-three minutes before the earthquake struck San Francisco, too late to warn its residents of

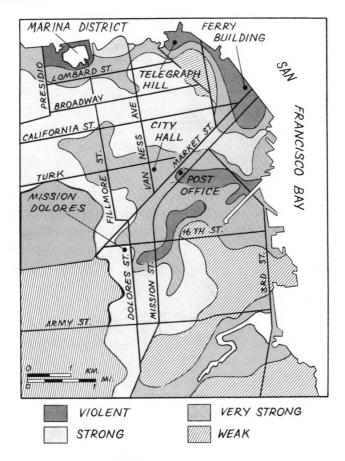

Fig. 9.5 Shaking Intensity from the 1906 San Francisco Earthquake

the impending disaster and to evacuate them from the danger zone.

Although John Milne had started installing seismographs throughout Japan in the 1880s, Alexander McAdie, San Francisco's weatherman in 1904, had just written Dr. F. Omori, Japan's pre-eminent seismologist, to inquire about the cost of a seismograph. (At the time, Japan was the only country with seismographs.) Dr. Omori responded that a seismograph could be obtained for about two thousand dollars and that it would give advance warning of an earthquake. But, by 1906, no seismograph had been either purchased or installed in San Francisco.

A 430 km (269 mi.) section of the San Andreas Fault ripped open at a speed of 3 km/sec. (1.9 mi./sec.). It started at a point deep in the Pacific Ocean and surfaced at Point Arena, 144 km (90 mi.) north of San Francisco, where it shattered the lantern of the lighthouse, before continuing southward through the city and ending at a settlement south of the city, where it smashed the Mission of San Juan Bautista (Fig. 9.6). On this occasion, the ground on the western side of the fault on the Pacific plate moved northward as much as 6.3 m (21 ft.), with respect to the North American plate.

Most people were still asleep when the quake woke them, but when gravestones were thrown to the ground in San Francisco's cemeteries, it was said that even the dead had been awakened. Such was the strength of the shock that the Valencia Hotel, a four-story wooden structure, collapsed into its own basement and the 800-room Palace Hotel, San Francisco's most luxurious, shed its rear wall, although it was so well built that, except for some cracks

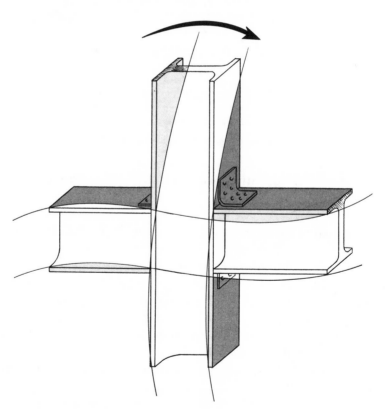

Fig. 9.6   Semi-Rigid Connection

in its front facade, it was otherwise undamaged. The Palace had been built in 1875 to be earthquake-proof and even had its own water supply to fight fires. It was *the* place to stay in San Francisco.

The world-famous tenor Enrico Caruso was sleeping soundly in his suite at the Palace, following a stirring performance in Bizet's *Carmen*, when he was shaken awake by the quake. He panicked and was found weeping hysterically by Alfred Hertz, the opera company's conductor. Caruso (like most Neopolitans, a believer in the evil eye), was convinced that *scarogna* (bad luck) had followed him to America, since eleven days earlier he had barely avoided an eruption of Mount Vesuvius, which had threatened Naples with a river of scalding lava. The singer even feared that the shock might have damaged his vocal chords, until Hertz distracted him from his neurotic fears and convinced him to open the window, look at the devastating scene and *sing*. Startling the people on the street, the great tenor sang: his magnificent voice rang out across the rubble-strewn streets and it was said that this was the singer's "bravest and best performance", showing that, contrary to fact, "at least *he* had not been scared".

The major shock, with an estimated magnitude of 8.25, lasted only forty seconds, but the shaking continued for almost ten minutes, doing the greatest damage to structures on *reclaimed* land (see Fig. 9.5). There, a deep fissure opened in the filled ground near the shore, pavements buckled, brick houses were severely damaged or totally destroyed and water and sewer lines were broken. At first, the only sound produced by the quake was a deep rumbling, but as the undulations of the ground set buildings into swaying motion, there came moaning and creaking sounds that built up to a roar when buildings began collapsing. Clouds of dust filled the air from crushed plaster and bricks and, added to the ominous sounds, created a terrible sense of confusion and fear in people all over the city.

Streetcar tracks were bent into wave-like forms, chimneys and brick walls crashed to the ground and church bells, moved by unseen hands, rang out. The masonry skin of the new city hall was peeled off and crumbled, leaving its ghostly steel frame standing, but the sixteen story Call building and the incomplete Fairmont Hotel on Nob Hill were only slightly damaged and survived until . . .

Below the streets of San Francisco, gas and water mains ruptured, sending geysers shooting up through broken pavement and

drowning some residents who had been pinned in wrecked buildings. Fifteen minutes after the quake subsided, fifty fires burned in
the business district. No alarms rang in the city's firehouses
because the central alarm system located in Chinatown had been
silenced when 93 percent of its emergency batteries shattered at
the first shock. The same shock had severed all the iron pipes from
the city's main storage reservoirs (that had survived) and broken
as many as twenty-three thousand underground pipes, preventing
the pressurization of the distribution system: deprived of water,
the city's firemen stood by helplessly as buildings burned. The Call
building and the Fairmont Hotel, both survivors of the quake, were
early victims of the fire. After the first flames erupted, winds, sometimes approaching gale force, blew uninterruptedly for three days
from the Pacific Ocean through the Golden Gate into the city.

Brigadier General Frederick Funston deployed his troops
throughout the city to forestall looting and to control the spreading fires, establishing "de facto" martial law. In an attempt to halt
the advancing wall of flames by creating firebreaks, Funston
ordered the demolition of a number of buildings, but the troops,
ill prepared to deal with such a disaster, created more damage
than necessary by blowing up burning buildings and causing red-
hot debris to fly in all directions. They aided rather than blocked
the approaching fires. The fires, finding little remaining consumable matter, finally died down: they had burned for seventy-four
hours and destroyed 12 km² (4.7 sq. mi.) of the city and twenty-
eight thousand buildings (Fig. 9.7). At the end of the third day, a
light rain doused the smoldering ashes. It is not surprising that
this tragic event was henceforth usually referred to as the great
San Francisco *fire*, rather than the San Francisco *earthquake*.

An exhausted population surveyed the ruins and counted its
losses: over 500 dead (6 shot by Funston's troops for looting or
other criminal activities), 5,000 injured and 100,000 homeless. As
the site of the first major earthquake of the twentieth century, San
Francisco demonstrated the strength of steel frames in resisting
the quake's destructive force: had it not been for the fire, the city
would have been damaged but not destroyed.

The seismic shock waves shook up even ships at sea near the
coastline of the Pacific Ocean: one of them, the *Argo*, lying 72 m
(240 ft.) directly above the fault, suffered a buckled steel-plate hull
and popped rivets from the shock. Yet, because the tectonic movement along the San Andreas fault is a *lateral* slip, the residents of

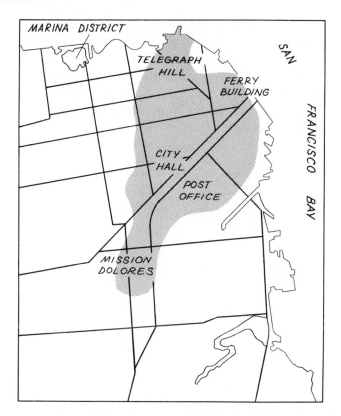

Fig. 9.7 Extent of Fire from the 1906 San Francisco Earthquake

San Francisco were spared the devastation of a tsunami that usually accompanies a vertical or *thrust* fault displacement: the sea only rippled in a mild vibratory motion.

In the fourteen months following the quake, a total of 153 aftershocks were recorded, a modest number when compared with the 1,256 aftershocks in the first month after the 1923 Tokyo earthquake.

In the haste to rebuild the city, rubble left in the wake of the disaster was pushed into San Francisco Bay, creating new land upon which to build. Once again, the lesson that should have been learned by the city from the 1868 quake was forgotten and buildings were again erected on "made ground" reclaimed from San Francisco Bay. In the Marina District (see Fig. 9.5), the lagoon was filled with sand to prepare for the 1912 Panama-Pacific International Exposition, a most unfortunate choice because, when wet, sand liquifies faster than any other soil during the shaking of an earthquake. The shortsightedness of all these actions became evident in 1989 when the moderate Loma Prieta earthquake of magnitude 7.1 struck the city, again devastating most buildings in the low-lying districts built on reclaimed land and causing the land in the Marina District to settle as much as 125 mm (5 in.).

10

Three Minutes in Anchorage

All things have second birth;
the earthquake is not satisfied at once.

—WILLIAM WORDSWORTH

Plate tectonic theory remained somewhat controversial until the day in 1964 when the earth roared so loudly that it silenced even its most skeptical opponents.

Valdez Harbor lies at the end of a 16 km (10 mi.) long fjord in the northwest corner of Prince William Sound in Alaska (Fig. 10.1). The 20,000-ton freighter *Chena* had arrived in the harbor earlier in the afternoon of Good Friday, March 27, 1964, and was tied up to the dock. Longshoremen were preparing to unload a bulldozer and other freight from its forward hold when, at precisely 25 seconds after 5:36 of that otherwise quiet afternoon, the big ship began to move, slamming madly against the wooden dock, and a deep rumble, like the sound of a passing train, ran through the ship, terrifying crew and officers alike. As Captain Merrill Stewart reached the bridge, the ship was rising above the dock and the heavy hemp ropes holding it were snapping apart. Townspeople who were standing on the dock to watch the unloading were running for

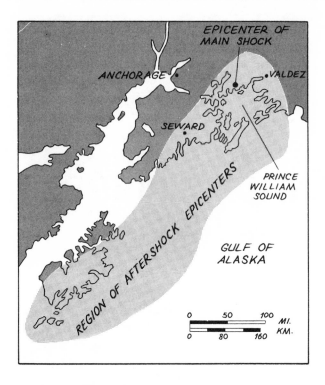

Fig. 10.1   The Gulf of Alaska

their lives toward shore as the dock began collapsing under them. Within a minute, the warehouses on the dock collapsed into the churning water and a number of people tried desperately to hold onto the sinking dock before disappearing into the frigid water. The *Chena* rolled to port, listing toward the dock at more than 45 degrees, dangerously close to capsizing as her stern rose high above the disintegrating dock before settling back into the water and slowly rolling to starboard. The big ship was being pushed around by the sea, which was sucked out of Valdez Harbor by a drop in the ocean floor and which then roared back as a tsunami. As the trough of the huge wave passed, the ship slammed into the bottom of the harbor (Fig. 10.2). Fearing that his ship would break apart from the repeated jarring impacts, Captain Stewart ordered the engines started to gain control of the ship, in the hope that he could steam out to the open sea. Just then, a second tsunami struck, pushing the ship dangerously close to shore and again rolling it sickeningly close to capsizing. But miraculously, the ship's

Fig. 10.2   The *Chena* in Port

propeller began slowly turning and the terrified crew were relieved when it sluggishly started moving away from the scene of the disaster. Its rolling motion continued, but slowly decreased as the ship snaked its way to the safety of the open sea through the debris-laden waters.

Meanwhile, the same seismic shock had rolled across the city of Anchorage, less than 190 km (120 mi.) away, wreaking havoc. North of the city, the elegant homes on the bluffs facing the Knick Arm Fjord slid down to the shore as a thick layer of "bootleggers cove clay", deep beneath their homes, liquefied from the shaking. The Turnagain Heights area on which the homes sat, about 300 m (1,000 ft.) wide and 3.2 km (2 mi.) long, broke into thousands of soil blocks, some of which dropped more than 15 m (50 ft.) to the bottom of the bluff (Fig. 10.3). Seventy-five homes were totally destroyed, while some danced convulsively down the cliff, landing on the bottom with their distorted wooden frames intact.

Witnesses to the Alaska quake were amazed at its duration: "I

Fig. 10.3   The Breakup of Turnagain Heights

had just entered my automobile when the earthquake started. . . . I was looking at the West Anchorage high school and kept thinking to myself: 'Well now, this earthquake has gone on long enough, now is the time for it to stop'. However, it continued and grew in intensity. Finally, after I had watched the school for some time, all of the glass in the second story seemed to explode and shattered all at once. The roof slab moved in waves that started from the two extreme ends of the classroom section and seemed to work toward the center and back again. This continued for some time. Then the roof slab seemed to sigh and rise slightly and, when it settled, all of the second-story columns broke". The shaking had lasted almost three minutes, although to some observers it was five eternal minutes. Surprisingly, small stiff structures of masonry construction survived, while many tall buildings, which have long periods of vibration, were severely damaged. This occured because of the predominance of long waves in the ground shaking. The seven-story-high, reinforced-concrete control tower at the airport was reduced to a pile of rubble. The six-story Four Seasons apartment house, constructed using the *lift-slab*[1] method, totally collapsed

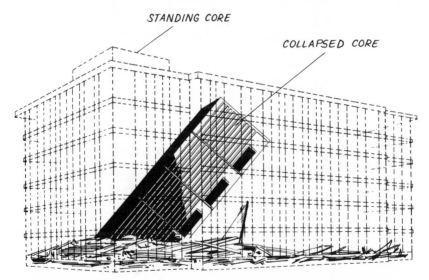

Fig. 10.4   The Collapse of the Four Seasons Apartment Building

[1] See *Why Buildings Fall Down*, by Matthys Levy and Mario Salvadori (W. W. Norton, 1992) for the story of the deadly collapse of the lift-slab-constructed L'Ambiance Plaza Apartment building. In lift-slab construction, concrete slabs are cast on top of each other on the ground like a stack of pancakes and lifted, one at a time, to their final elevation.

from the persistent shaking (Fig. 10.4). Several other tall apartment buildings were severely damaged, many exhibiting cross-shaped cracks in their masonry spandrels between windows, evidence of severe back-and-forth bending of the building (Fig. 10.5). Oil storage tanks were severely buckled by the sideward push of liquid as their contents sloshed back and forth. (Fig. 10.6).

The Alaska earthquake of 1964 was rated magnitude 8.4, the largest ever recorded in North America. The epicenter of the main shock was located between Valdez and Anchorage on a *thrust fault* almost 960 km (600 mi.) long where decades of accumulated strain had been suddenly released along the edge between the giant Pacific plate pushing northwestward under the North American plate. The land on the southeast side of the plate boundary was lifted as much as 11.4 m (38 ft.), while the northwest side dropped as much as 2.3 m (7.5 ft.), and the two sides of the boundary slid horizontally against each other 19 m (64 ft.). The violent shaking

Fig. 10.5   Cross-Shaped Spandrel Cracking

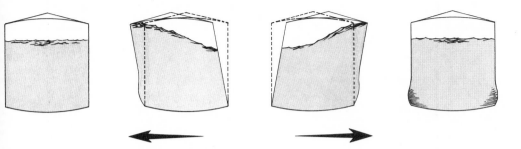

Fig. 10.6   Sloshing of Water and Buckling of a Tank in an Earthquake

opened fissures in the ground, and caused rockslides and mud spouts in a large area of southwestern Alaska. In the months following the quake, of more than twelve thousand recorded aftershocks, nineteen were of magnitude 6 or above.

The impact of the quake was so extensive that the central part of the United States rose imperceptibly 75 mm (3 in.) and seiche action in rivers, lakes and harbors was reported to cause minor damage as far away as the gulf coast of Lousiana. Even the Caribbean island of Cuba, more than 6 400 km (4,000 mi.) away, shook. The quake deformed a larger area (250 000 km², [100,000 sq. mi.]) of the earth's surface than any other quake in historic time. The tsunami resulting from the uplifting of the seabed by 15 m (50 ft.) was so powerful that it compressed the earth's atmosphere, causing ripples in the ionosphere 90 km (50 mi.) above the surface of the earth.

The low loss of life (131 died, of whom 122 perished in the tsunami) can only be attributed to the sparsity of population in Alaska as well as the time of the earthquake's occurrence, after the end of the school and business day, and at low tide, preventing greater damage from wave action. But the tsunami not only devastated towns on the Gulf of Alaska, it also caused flooding and drownings as it landed on the beaches of the western United States and the islands of Hawaii.

II

Precursors and Predictions

I remember when our island was shaken with an earthquake some years ago, there was an impudent mountebank who sold pills which (as he told the country people) were very good against an earthquake.

—JOSEPH ADDISON

As far back as recorded earthquake history in North America, a period of about 1,500 years, a major earthquake has occurred about every 150 years along the southern portion of California's San Andreas Fault (see p. 100), the last one in 1857. It is therefore considered probable by the seismologists that a major quake will shake that particular area in the decade shortly before or after the year 2000. Luckily, it will not come unannounced: foreshocks of about magnitude 5 will increase in frequency over several years; the ground may rise in the region and measured distances, say along property lines, may change in length; magnetic or gravitational anomalies may be recorded and strainmeters (devices that measure the relative motion of two adjacent points on the ground), may indicate additional changes on the earth's surface. But they could warn us only if we could read all these earthquake precursors with great accuracy: unfortunately, we are not there yet.

Long-Term Predictions

In 1992, tree-ring analysis was able to prove that long-buried Douglas firs, drowned when carried by a landslide into Seattle's Lake Washington, and a single log, felled by a tsunami 22 km (14 mi.) away, on the shores of Puget Sound, died in the same season of the same year. Scientists determined that the log from Puget Sound, discovered under sand on the shore, was deposited by a tsunami. They used radiocarbon dating to approximate the tree's death at between 1,000 and 1,100 years ago. Tree-ring analysis of the firs in Lake Washington proved that the landslide had occurred in the same time span. Hence, the two events were simultaneous and were probably caused by a strong earthquake that was dated—again, by tree-ring analysis—at about A.D. 992. The firs in Lake Washington thus confirmed that Seattle is in an earthquake-prone zone and alerted seismologists to a novel approach to seismic dating, providing another tool for long-term predictions.

The San Andreas Fault is perhaps the most "read" fault in the world. Thousands of earthquakes of magnitude 1.5 or higher strike the region along the San Andreas every year and seismologists have stationed some seven hundred seismometers in the fault region to record the response to even the weakest quake. Geologists, by cutting trenches across sections of the fault, have recently identified historic and prehistoric earthquakes, by noting the discontinuous shifts in the soil layers that result from a seismic event (Fig. 11.1). As the stratified faces of the Grand Canyon have so clearly demonstrated, these cross sections of the earth provide a visual record of historic events: samples taken from different depths are radiocarbon-dated to pinpoint the date of the event, whether earthquake, volcanic eruption, drought or flood. As a result, a new discipline, *paleoseismology*, is contributing information to the prediction of earthquakes by **extrapolation** of the recurrence frequency of past earthquakes. Using both the historic and current seismic data, seismologists have, within the last twenty years, put together a picture of past earthquakes and can now speculate on the future seismicity of a region.

The seismologists first divided the 1 300 km (800 mi.) length of the San Andreas Fault region into sections with similar characteristics, such as geology and rate of movement (strain rate) along the fault. They then assigned to each section a given probability that

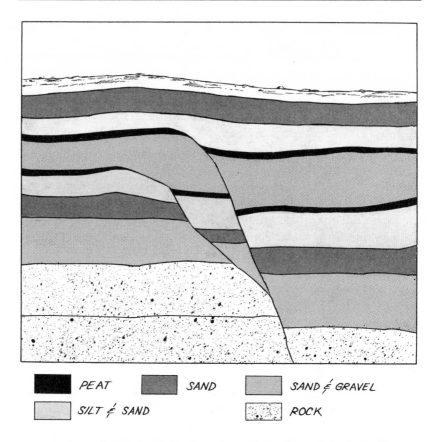

PEAT   SAND   SAND & GRAVEL

SILT & SAND   ROCK

Fig. 11.1   Shifting of Underground Soils in Earthquakes

a major earthquake would occur in the next thirty years, based on past history and rate of activity (Fig. 11.2). Under this scenario, the region around San Bernardino has a 60 percent risk, the San Francisco area a 67 percent risk and the Parkfield area in west central California, a 90 percent risk of having a major earthquake in the next thirty years. Using this method, seismologists had correctly assigned the *highest* risk to the Santa Cruz Mountains near Loma Prieta, where the 1989 earthquake occurred.

The Parkfield area was assigned a high seismic probability because of the unique periodic regularity of its earthquakes: a major earthquake of magnitude 5 or 6 strikes the area, on average, every twenty-two years (only once did as many as thirty-two years elapse between two quakes). Parkfield is now saturated with instruments (seismometers, accelerometers, creepmeters, strain-

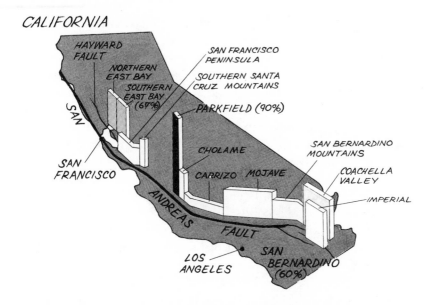

Fig. 11.2   Earthquake Probabilities along the San Andreas Fault

meters, tiltmeters, magnetometers, geochemical sensors and laser levels) that record the array of data needed to discover the meaning of the precursors and to pinpoint which data can be used as accurate predictors of earthquakes.

All predictions based on recurrence intervals or strain rates are necessarily long-term since, at best, they can pinpoint the place and the magnitude of an earthquake but they can establish the time of occurence only within about thirty years, or a human generation. Yet, once in a while, nature provides particular hints and may even change the rules of the game, as happened at Yucca Valley in 1992.

Precursors to the Big One

Yucca Valley, 36 km (22 mi.) northeast of Palm Springs, California, and 200 km (125 mi.) east of Los Angeles, is the refuge for people calling themselves "desert rats" who are tired of life in the big city. They come to this hot, dusty area from all walks of life—retired professionals, young families, postal service employees, flying instructors and the storekeepers of the establishments lining the road along the valley. They all love the relative solitude, the clear skies and the wide open spaces of the area.

But there is no guarantee of peace even under the shade of the yucca trees, with their small leaves and white flower tufts (Fig. 11.3). At 4:58 A.M. on June 28, 1992, an earthquake of magnitude 7.5, the strongest in forty years in California and five times more powerful than the 1989 Loma Prieta quake, brought havoc to the town of Landers, a few kilometers northeast of the epicenter. The quake destroyed twenty houses and ten businesses, burned four houses to the ground, damaged one thousand houses and injured three hundred people badly enough to require medical treatment. It lasted sixty eternal seconds.

Later that same morning, residents of an area near Big Bear City, 26 km (16 mi.) northwest of Landers in the San Bernardino Mountains, were scared out of their wits at 8:05 A.M., by a magnitude 6.6 shock (Fig. 11.4). Fifty-four aftershocks of a magnitude up to 5.4 followed within a few hours. Had the Landers and Big Bear City quakes occurred in the Los Angeles area, at least ten thousand people might have died. Instead, the "only" casualty was a toddler killed in his sleep by a falling cinder block, a tragedy for his par-

Fig. 11.3   The Yucca Valley, 1992

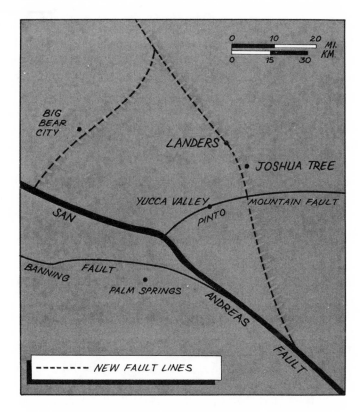

Fig. 11.4   Yucca Valley Fault Triangle

ents, who were in Landers from New Jersey for a nostalgic high school reunion.

Yet, despite the destruction and the pain, no inhabitant of the valley intends to leave. Their justifications are varied and many but are all based on a deeply ingrained fatalism and an abstract comparison between various kinds of natural disasters. "Are volcanic eruptions or hurricanes or tsunamis preferable?" "Is life in Los Angeles protected by a handgun in your pocket?" "I prefer tornadoes any time: at least they give warning, but . . ." "I really do not believe that I am going to die in an earthquake."

Our attachment to the land, which the Italians call the love of the *patria piccola* (the little motherland), takes many aspects. Thanks to our first astronauts, some of us consider our "little" motherland to be the whole earth, while others choose their country of birth or of choice, and others the land where their language or only their dialect is spoken. Many of us have a particular attachment to the small area where we were born or have seen our children grow. Hence, we should not be surprised that most people refuse to abandon their piece of land even when they know it is unwise to live there.

But the seismologists think they are wrong. The two quakes in the Yucca Valley were felt all the way to Washington State and Utah, and lifted the earth up to 5.4 m (18 ft.), breaking its crust. But more significantly, 24 km (15 mi.) below the earth's surface, *two* active faults, whose existence had never been suspected in that "safe" area, were discovered. One joins the San Andreas fault southeast of Big Bear City and the other joins it near Joshua Tree (the site of a magnitude 6.3 quake two months earlier), creating a new triangle of faults in southern California (see Fig. 11.4).

The series of earthquakes along two sides of this triangle have not relieved the accumulated strain along the third side, the San Andreas Fault. But the quakes caused the triangle to move about 1 m (3 ft.) northward, and reduced the pressure against the San Andreas Fault and made it easier for that section to suddenly slip (Fig. 11.5). The last anxious question is: When might a slip occur? The increased activity along this section of the fault (seven quakes since 1986), is similar to the pre-1906 activity around San Francisco and has led the young paleoseismologist Kerry E. Sieh to predict: "It's just a feeling, but I think I'll witness a great earthquake on the southern San Andreas in my lifetime."

Short-Term Predictions

A promising technique for foretelling earthquakes dates from the early 1960s, when Russian seismologists observed that the ratio of the travel time of the P-waves to that of the S waves (see p. 68) decreased over time only to return to its normal value just prior to a quake. Generally, P-waves travel 1.75 times faster than S waves, and for a period of time before an earthquake, P-waves slow down by 10 to 15 percent, possibly due to the *dilatancy* (swelling) of rock before it breaks. By analyzing data from prior earthquakes, these scientists also discovered that the *duration* of this temporary anomaly in the travel time ratio of the P-waves and S waves was proportional to the *magnitude* of the eventual earthquake. In 1973, the data accumulated from measurements of travel-time ratios

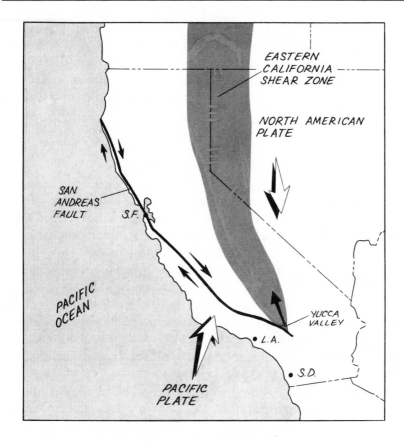

Fig. 11.5   Fault Movements in California

provided by a swarm of precursors in the Adirondack Mountains, at Blue Mountain Lake, New York, allowed scientists from the Lamont Doherty Earth Observatory to predict—for the first time—the location, magnitude *and* time of occurence of an earthquake. They proved to be quite accurate on all three predictions: a magnitude 2.6 quake occurred at Blue Mountain Lake within a day of the predicted date.

Another potential short-term predictor is based on the assumption that the sudden slip between two tectonic plates is facilitated by the lubricating effect of water at high pressure forcing its way into the joint between the plates. Based on this assumption, it follows that a surge of underground water, indicated by higher water levels in wells, could predict a seismic event.

It is hoped that analysis of the records from the experiments

placed around Parkfield will clarify the nature of earthquake precursors and permit a more accurate determination of *when* the
next earthquake will arrive. However, to date, no short-term predictive indicator has proven to be universally reliable, except one
that springs not from scientific observation but from many popular folklore tales.

The Cock's Crow

The popular tradition that unusual animal behavior is a good
earthquake predictor is as old as humanity: Chinese historians
have reported anecdotal episodes of strange animal behavior preceding earthquakes in chronicles over two thousand years old and
the first European (Greek) record of such behavior goes back to
373 B.C.

Despite the similarity of these stories from all parts of the
world, modern scientists have been skeptical of these anecdotes,
which lack convincing scientific explanations. Even so, whatever
their geographical and historical origins, traditional stories about
strange phenomena preceding recent and historical earthquakes
are so similar that they have awakened a renewed interest in their
predictive capacity by Chinese, Japanese and, more recently, even
U.S. seismologists.

The earthquake folklore is almost unanimous in attributing
the cause of earthquakes to a variety of animals belonging to the
tradition of the countries involved. In southern Chile a legend
describes how earthquakes are due to the fights between two big
snakes, Cai-cay, mistress of the waters, and Tre-treg, who fills with
stones the holes Cai-cay digs to store her waters. Snakes responsible for earth tremors appear in the Greco-Roman tradition and in
early medieval Italian palimpsests. In ancient China the winged
dragon-snake Lung shook the earth and in India an elephant was
believed to be the cause of seismic tremors. A giant bull shook the
earth on its horns in the Caucasus and the Hebrews made a bull
responsible for their seismic catastrophes, just as the bull of
Knossos shook the island of Crete. In Kamchatka, a peninsula off
the Siberian coast, earthquakes were believed to be caused by a
subterranean dog and in Mexico by a jaguar. A subterranean frog
or fish caused the local earthquakes in central Asia and an under-
world goddess was the originator of earthquakes in Babylon.

The most thorough compilation of these traditions and anecdotal stories is due to Dr. Helmut Tributsch,[1] a German scientist
and native of the province of Friuli in northern Italy. Dr. Tributsch
was motivated by the magnitude 6.5 earthquake that occurred in
Friuli at 9 A.M. on May 6, 1976. It killed 1,000 people and destroyed
100,000 houses, including his parents' ancestral home, leading Dr.
Tributsch to dedicate his life to the study of the folkloric phenomena preceding earthquakes described by local witnesses.

Dr. Tributsch assembled evidence of 178 episodes of unusual
animal behavior in Europe, China, Japan, the Americas and other
parts of the world and he suggested a scientific explanation of
physical phenomena, as well as of human physiological states, that
supposedly precede earthquakes. According to eyewitness
accounts, the following partial list of animal and human behaviors
and physical phenomena are said to precede earthquakes by as
much as a month and a half, and by as little as a few seconds:

- Four-legged animals show nervousness and break out of their
  stables or, on the contrary, refuse to move;
- Flocks of birds suddenly fly in circles or fly away at great
  speed;
- All dogs in a given neighborhood howl for entire days and
  nights;
- Roosters fly to the top of trees and refuse to abandon their
  perches;
- Wild animals, like tigers, behave tamely;
- Flies and mosquitoes disappear suddenly from sites they
  usually infest;
- Snakes come out of hibernation and freeze on snow;
- Wintering bears come out of their lairs;
- All the cats run out of the houses of an entire village;
- Deep-sea fish appear on the sea surface and die on beaches;
- Surface fish jump out of the water in incredible numbers and
  some land on beaches;
- Mice stop running, frozen by fear, and can be easily grabbed,
  even by children;
- Waters in springs and lakes become suddenly muddied;
- Water tables suddenly change level;

[1] See Dr. Tributsch's book *When the Snakes Awake* (Cambridge: MIT Press,
1982), an English translation by Paul Langner of the original 1978 German edition.

- A variety of roaring sounds are heard emanating from the earth;
- Lights appear in the sky (some have been photographed in color);
- Flowers bloom ahead of their season;
- Humans feel unusually weak; and
- Humans predict earthquakes when the air becomes unusually hot and dry ("earthquake weather").

Once traditional precursors are scientifically explained, they could help in the early prediction of earthquakes and save a large number of lives. Despite the lack of unanimous acceptance by seismologists, over the last decades, the People's Republic of China has organized large numbers of farmers, students, teachers, soldiers and seismologists (as many as 100,000) in a network of observers of such presumed earthquake precursors. Their observations are immediately transmitted to China's seismological centers and analyzed by seismologists, who, when deemed appropriate, broadcast warnings to the supposedly endangered areas. Following months of increasing seismic activity in the Haicheng area, local networks became aware of the possibility of a major earthquake. Warnings based on the networks's reports were broadcast days before the February 4, 1975, earthquake of magnitude 7.3, and were credited with saving the lives of many people who, believing the forecasts, had left their homes. But before the Tangshan earthquake of July 28, 1976, which killed over 300,000 people (the largest historical death toll) there was no increased seismic activity and no heightened caution by the networks: no forecast was issued, increasing skepticism among the world's scientists about traditional earthquake prognostications.

But at the time of this writing (1994), enough interest about traditional precursors has developed in Japan and the United States to suggest the implementation of small networks of observers who will transmit their information to seismological centers for scientific study and for use as warnings.

Whether one is ready to accept some or all the conclusions of Dr. Tributsch, we do not believe that one can deny the reality of the ancient traditions connecting the unusual behavior of humans and animals that precedes phenomena as complex as earthquakes. Neither can we forget that many animals are endowed with senses more refined than our own. If birds, and even butterflies, can migrate enormous distances and return to their birthplaces, thanks to their sensitivity to the earth's magnetic field or the polarization of the sun's light; if the swallows punctually return to Capistrano (except when they sense an earthquake!); if dogs, using their acute sense of smell, can follow the traces of a human being; if bats can navigate in the dark by relying on their radar capabilities (emitting and receiving high-frequency ultrasounds); and if salmon, after navigating the oceans, come back to breed and die at the head of the rivers where they were born thanks to their incredible sense of smell, it is hard *not* to believe that animals may be more sensitive than we are to changes in magnetic or gravity fields, to infra- and ultra-sounds and to all the other phenomena that we know accompany or precede earthquakes.

12

Seismic Isolation

> They entered the forecourt . . . thunder and
> lightning struck down the Persians and
> then from close by Mount Parnassos, two
> great rocks fell and crushed the Persians.
>
> —KONRAD WOLFHART

First There Was Damping

When, toward the end of the eighteenth century, human societies
began to consider earthquakes as natural phenomena rather than
God's retribution for their sins, technologists started to look for
ways and means to avoid, or at least reduce, the damage caused
by the sudden, as yet unexplained movements of the earth's crust.
But for two more centuries they seemed to stubbornly follow a
popular philosophy that dictated: "If you can't fight them, join
them." Paraphrasing the word of the American architect Christo-
pher Arnold, they constructed buildings as strong as economy
would allow and attached them securely to the ground, "trying to
arm wrestle with nature".

When the earth shakes, a large amount of energy goes into and
must be absorbed by a building structure founded on the earth's
crust. Designers have long recognized that the more this energy

can be dissipated elsewhere, the less is the strength that has to be built into structures to resist seismic events. We have already described how the more rigid a building is, the harder it is shaken by the earthquake, since it follows the motions of the ground and, hence, the more likely that it will be damaged or destroyed, as rigid masonry buildings have been in every major earthquake. But a flexible structure that bends elastically rather than standing rigidly, and consequently transforms some of the earthquake's energy into bending energy, has a greater chance of survival, as amply demonstrated in recent earthquakes by the undamaged, flexible steel-framed buildings. Yet if a building is *too* flexible, it will sway uncomfortably and become uninhabitable, so that modern antiseismic buildings are a compromise between being stiff enough to be usable and flexible enough to resist seismic forces.

There are other ways to absorb energy besides deforming the building's elastic structure. Most methods rely on introducing *dampers* into the structure's joints or its bracing elements. Dampers, like a car's shock absorbers, are energy-dissipating mechanisms that use friction, fluid-activated pistons or *viscoelastic* devices (they combine the actions of a spring and a fluid-activated piston), to accomplish their task.

Base Isolation

Of course, it would be ideal to allow a building to stand passively in an earthquake while remaining intact and unmoved. But is this feasible? We will probably never know the name of the genius who first thought of proposing a radical answer to this question: disconnect the building from the earth, letting the earth shake under it. If this were possible, the building would remain, more or less, unmoved, as a consequence of Newton's *law of inertia*, which asserts that a body at rest does not move unless a force is applied to it. By the early twentieth century, this obvious and practical idea caught the imagination of inventors and amateurs alike. In 1907, Jacob Berchtold was granted a patent based on this principle and in 1909 J. A. Calantarients, an English medical doctor, published a paper entitled "Improvements in and Connected with Building and Other Works and Appurtenances to Resist the Action of Earthquakes and the Like." Calantarients assumed that earthquake movements are mostly horizontal (as they usually are) and suggested that a layer of talc between the bottom of a building and its foundation (resting on and bound with the underlying soil) would lubricate the surface of the joint and allow the foundation to slide back and forth *under* the building, which would stand *unmoved.*

Unfortunately, the time that passes between the suggestion of a sound, ingenious idea and its practical implementation is often surprisingly long: it was only toward the end of our century that the theoretical concept of *base isolation* was translated into the reality of its successful but complex implementation. Let us find out the reasons for this long delay.

The Requirements
of a Base Isolation System

Consider performing the following simple experiment: lay a toy car or a skate on a cardboard sheet and yank the cardboard back and forth in imitation of the horizontal motions of an earthquake. Notice that the car will not slide as much as the cardboard, but it will still move slightly back and forth and, at the end of the "earthquake", rest at a point slightly removed from the point it started at. This elementary experiment shows that, since we cannot allow an isolated building to be permanently displaced from its initial location, nor allow it to move back and forth by more than a few centimeters, its motions must be "damped out" and the building brought back to its original location.

To achieve these goals, we must invent one mechanism that will allow the building base to move and another that will limit its oscillations with respect to the earth. Such mechanisms will be effective only if the shaky motions of the earth they must respond to are known and if the inertial forces on the building due to the building's damped motions can be accurately estimated. In fact, until a few decades ago, neither of these two essential pieces of information nor the materials needed to allow the damped motions of the buildings' supports were available. Additionally, the science of structural dynamics, which determines the reactions of a building to earthquake motions, could not accurately determine the values of the inertial forces on a building until computer programs became available to solve these daunting mathematical problems.

Finally, the practical development of seismic isolation had to wait for accurate measurements of earthquake motions at the site of the building to be isolated and, unfortunately, earthquakes did not occur on demand to satisfy the need for such data. It might be thought that theoretical calculations could have supplied the necessary information, but since building developments involve human lives as well as large amounts of money, building officials are not inclined to accept conclusions based on purely theoretical investigations: they require tests. Hence, tests had to be devised capable of reproducing exactly the recorded motions of past earthquakes and the inertial forces they generated in the full-scale building. The solution was to test scale models of buildings on *shaking tables* that could reproduce the effects of an earthquake.

At long last, in the 1980s, all these requirements became available: seismographs provided accurate records of earthquake motion; shaking tables duplicated the actual earthquake motions and imposed them on models; computer programs easily solved the nonlinear problems of structural dynamics; and *elastomeric materials* of natural or artificial rubber made possible the manufacture of flexible pads allowing the horizontal displacement of the supports. Today hundreds of buildings, bridges and special structures, such as nuclear reactors, have been successfully isolated and new isolation techniques are being developed all over the world.

Seismic base isolation is a not a panacea, solving *all* our earthquake problems. For example, tall buildings that generally have long periods of oscillation (above three seconds) cannot be isolated with available elastomeric materials that do not damp out the ground's most dangerous long-period oscillations. Similarly, buildings on soft soils that filter out short-period motions cannot be isolated, since isolation lengthens the period of a stiff structure and consequently amplifies the effects of the long-period ground motions already felt by the structure. This was the case, for instance, in Mexico City, which was struck by a devastating earthquake in 1985, when the underlying seismic motion was dominated by long period components due to weak soils of the lake bed on which Mexico City rests.

Taming the Earthquakes

Elastomeric pads that allow the horizontal motions of the earth under the building while limiting those of the building, must also support the weight of the building without appreciable vertical displacements. To do so, the pads are made of alternating layers of rubber and steel plates that, while easily sliding in horizontal "shear" deflections, are vertically rigid (Fig. 12.1). The elasticity of the elastomerics, while permitting the horizontal (relative) motions between the building and the earth, must also limit the motions of the building to a few centimeters in order not to damage the expensive flexible connections of electrical and mechanical systems components crossing the level of the pads.

The layman may easily understand the efficacy of base isolation by examining the shape of a curve that plays an important role in seismic design: the *acceleration response spectrum* of an earthquake. Since all buildings vibrate under the impact of an earthquake, oscillating like upside-down pendulums (Fig. 12.2) with a basic period (the time it takes for a complete oscillation), acceleration response spectra are obtained by recording the response of simple masses on vertical blade springs of increasing length (Fig. 12.3a), equivalent to upside-down pendulums with increasing periods. The acceleration response spectrum is the plot of the maximum acceleration of the pendulums versus their increasing periods (Fig. 12.3b). As the figure shows, after an initial increase, the maximum accelerations *decrease* with *increasing* periods. Since the inertial forces on the building are in proportion to their accelerations and increasing periods indicate greater flexibilities, the spectra prove quantitatively that earthquake forces are smaller for flexible than for rigid buildings.

Evidence of this phenomenon was dramatically given by the behavior of the high-rise buildings in San Francisco during the

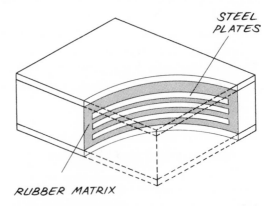

Fig. 12.1   Elastomeric Pads

Fig. 12.2   Oscillations of an Inverted Pendulum

magnitude 7.1 Loma Prieta earthquake of 1989. While older, stiff, unreinforced masonry buildings were seriously damaged or collapsed, flexible high-rises designed in conformity with recent seismic codes waved like trees in the wind but were entirely undamaged. The reader may have seen an identical behavior when a violent storm hits a forest: an old, stiff, strong-looking oak may be cut down by the wind, while a young, flexible willow bends and springs back undamaged.

Damping the Motions of Isolated Systems

Once we isolate the structure of a building from the earth, we must make sure that the building does not move excessively and, above all, that it will go back to its original location.

The first prerequisite may be obtained in a variety of ways, but perhaps the simplest is to mix into the rubber of the pads special-purpose fillers that increase the pads' *internal friction:* the back-and-forth motion of the building relative to the earth is thus damped. Moreover, friction absorbs some of the earthquake energy that would otherwise go into shaking the building and so reduces the quake's impact on the structure. (Remember that energy *never* disappears completely but only changes from one form to another

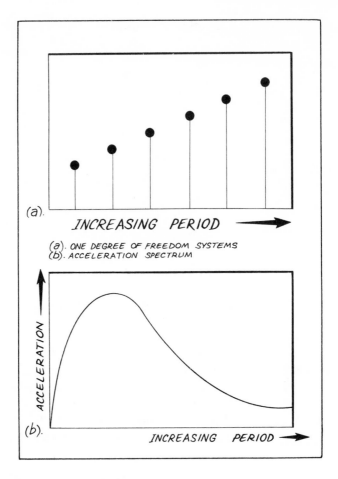

Fig. 12.3   Acceleration Spectrum

and that energy absorbed by friction is transformed into harmless heat.) Damped elastomeric pads offer the additional advantage of preventing, with their frictional resistance, the frequent and annoying small oscillations that even a light wind or vehicular movement would excite in a building or bridge supported on perfectly elastic pads.

Another device commonly used to dampen building motions consists of inserting at the center of each elastomeric pad a cylinder of lead, that increases substantially the amount of energy dissipated in the *shearing* distortion of the pad thanks to the plasticity of lead (Fig. 12.4).

In all isolation systems the final problem of returning the

building to its original location is usually solved by means of the action of springs or the pull of horizontal steel bars acting as springs (Fig. 12.5) or by the use of elastomeric pads with vertical steel rods in the center that act as vertical cantilever springs, thanks to the elasticity of the steel rods.

The Weak First Floor Isolator

Long before elastomeric pads were invented, a serendipitous method of base isolation appeared accidentaly in three- to four-story townhouses built in U.S. cities during the first decade of the century. These townhouses had a ground floor entirely dedicated to garages (Fig. 12.6). In an earthquake, the columns of the ground stories, unbraced by walls, bend under the inertial forces of the heavy, stiff floors they support and are often shifted precariously out of plumb while the upper floors remain intact. Although returning such a building to its original shape is sometimes impossible and often leads to its demolition, it suggests another method of building isolation: *the weak first floor system*. It consists in hinging both ends of basement columns and restraining the ground floor with springs (Fig. 12.7) or in placing piles in pipe sleeves of a slightly larger diameter, hinging them at the top and bottom and then connecting the tops to ground-level steel bars encased in concrete blocks (Fig. 12.8). In the second case, the pipe sleeves limit the displacement of the building, while the steel bars connected to the heavy concrete blocks act as springs and return the building to its original location.

According to Frank Lloyd Wright, the Imperial Hotel was

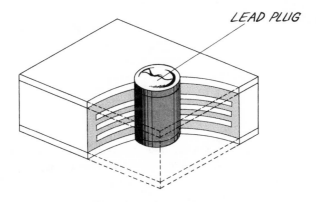

Fig. 12.4   Damped Elastomeric Pad

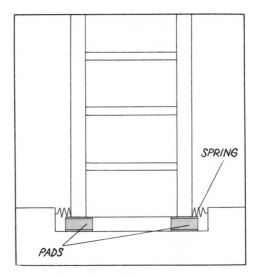

Fig. 12.5   Base-Isolated Building Restrained by Springs

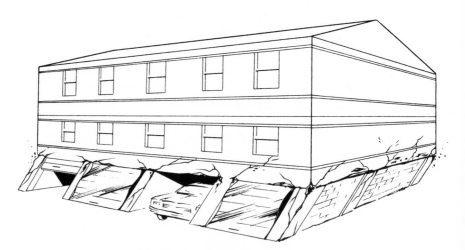

Fig. 12.6   Weak First-Floor Distortion

saved from the disastrous 1923 Tokyo earthquake by its long, flexible foundation piles that, acting as the underground columns of a "weak first floor", allowed the heavy reinforced concrete building to oscillate horizontally at their top without damage.

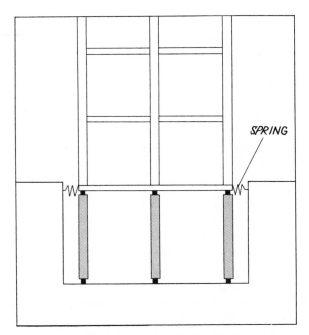

Fig. 12.7   Hinged Post-Supported Building Restrained by Springs

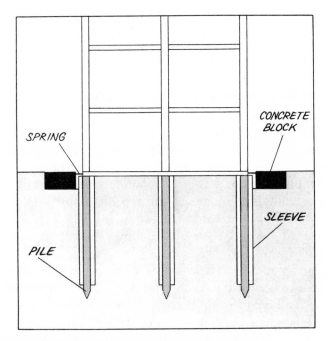

Fig. 12.8   Sleeved-Pile Foundation

Isolation Costs

Like any other antiseismic remedy, isolation must prove economical before becoming accepted. The larger the motions permitted by the isolation system, the smaller the earthquake forces on the building and, hence, the larger the savings in the cost of the structure. But large displacements increase the cost of flexible electrical and mechanical systems at the level of the isolation system, where differential horizontal displacements must be allowed to occur. Similarly, large damping in the isolation system shortens the period of oscillations of the building, but increases the earthquake forces on it, indicating that isolation requires an "economic balance" as well as a purely dynamic balance.

Notably, the Getty Museum in Los Angeles has adopted base isolation for many of its most precious pieces of sculpture, like the fifth-century B.C. Aphrodite. The statue rests on an isolator that permits it to slide gently back and forth on its marble base, unharmed by wave after wave of earthquake jolts. The cost of these isolators is a tiny fraction of the value of such priceless art. It is also noteworthy that the new California requirements for the isolation of nonstructural elements in hospitals add less than 1 percent to the cost of a comparable unisolated hospital.

A New Zealand Retrofit

The Union House is a twelve-story prefabricated concrete office building in Auckland, New Zealand, a country with earthquake problems equivalent to those in California. Union House is built on reclaimed land requiring 9 to 12 m (30 to 40 ft.) long piles. To isolate the nonductile concrete structure of the building (Fig. 12.9), the concrete piles have been placed in steel sleeves and weakened at the top and bottom, thus creating hinges that allow horizontal motions by acting as an "underground weak first story" (see Fig. 12.8). The motion at the top of the piles is limited by *energy dissipators* consisting of steel bars in concrete blocks, which, when pulled beyond their elastic limit by strong earthquakes, require replacement.

The framed superstructure is cross-braced by steel plates embedded in concrete to reduce its period, and the elevator shaft is a non-load-bearing, fireproofed wood structure supported on a steel space frame that is hung into elevator pits from the second floor. This pendulum-like structure has a clearance of

Fig. 12.9   The Union House Isolated Building

150 mm (6 in.) with respect to the first floor and basement.

Because of the speed of the prefabricated construction, which cut two months from the construction schedule, and the simplicity of its superstructure, the Union House was built with savings of about 4 percent of its anticipated cost.

No earthquake has tested the system since the time of its first description in the press (1989), but theoretical calculations estimated that, in the event of a shock of a magnitude expected in the area, damage would be negligible.

A Retrofit in California

The six hundred employees of the Rockwell International Company of Seal Beach, California, are in charge of monitoring all the NASA space shuttle launches and flights. The corporation building sits 1 km (0.7 mi.) from the Newport-Inglewood Fault and an expected shock of magnitude 7.0. The Rockwell operation cannot afford to be interrupted *at any time* and a thorough *retrofit* (structural modification) of the building was completed in 1992 to guarantee its integrity and that of the delicate instruments it contains.

The eight-story, concrete frame/floor-slab structure, built in 1960 for office purposes, was retrofitted with the construction of a partial, reinforced concrete outer frame (Fig. 12.10), satisfying the ductility requirements of the most recent code *and* with an isolation system of elastomeric pads, skirting a stairwell and the elevator core, and a pendulum structure for the elevator cores. The elastomeric supports are single rubber/steel-laminated units under the twenty-six interior columns; double units with damping lead cores under the twenty-four intermediate outer columns; and single pads with lead cores under the four corner columns, all set between the first and second floors. The lead cores, whose main purpose is to stop even minor vibrations due to wind, are of two sizes to compensate for the different torsional structural resistances of the different parts of the building. The outer columns, supported by double-pad units, carry 90 percent of the seismic load and have been structurally strengthened.

The delicate job of setting the elastomeric pads between the first and second floors of the building required cutting a chunk of column at a time, supporting the weight of the upper seven floors on a temporary jack, inserting the pad and then lowering the column's upper part onto the pad. But by far the most complex job of the retrofit consisted in cutting the walls of the two elevator cores to make them into pendulums *without* affecting the operational capacity of the building during an earthquake. (Elastomeric pads could not be used to isolate the cores because differential horizontal movements of the cores with respect to the building were

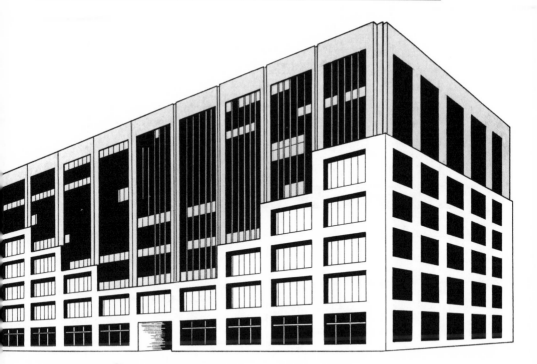

Fig. 12.10 Rockwell Retrofit

expected to reach 360 mm [14 in.]) Hence, the designers decided to sever the elevator cores at the footing level and to insert in the cut lubricating plates of *Teflon* that provide neither lateral support nor damping, but which allow the cores to move together with the upper part of the building as a gigantic pendulum hanging from the upper structure. To finish the retrofit, all utility lines had to be made flexible where they pass through the isolation plane.

But, most important, the code authorities had to be convinced, through a six-month period of meetings and discussions plus the time prior to submittal of the project, that the building, having been designed with a factor of two over the requirements of the code, would be safe. Engineering skills cannot be limited to purely technological knowledge: they must include unusual powers of persuasion whenever a new system is introduced in the design of a structure on the basis of purely mathematical considerations, a situation at times hard to accept by the savvy hard-hat engineer who has the final responsibility for the construction of a complex building.

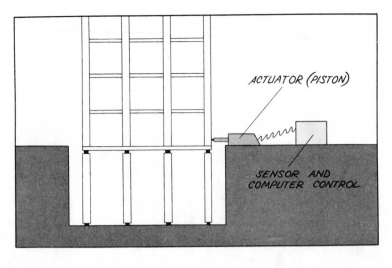

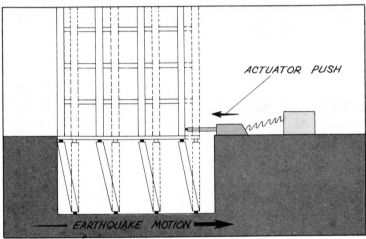

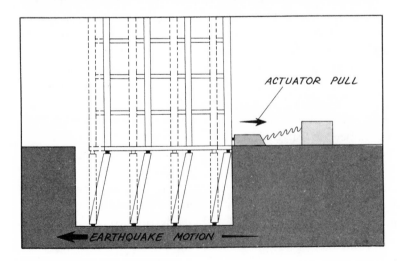

Fig. 12.11   Active Controls

Active Seismic Controls

The availability of high-speed, low-cost computers has led to the development of *computer-controlled* antiseismic systems.

Whereas the flexibility and damping devices of the isolation systems we have mentioned so far are *passive*—that is, they respond automatically to the seismic inertial forces—*active* control systems are driven by *mechanical actuators* that are controlled by electronic devices that "sense" the accelerations of the earth's crust due to an earthquake. The mechanical actuators are instructed by the sensing devices to impose on a building forces *equal but opposite* to those due to the earthquake, thus effectively neutralizing the earthquake's impact on the building.

In an application of an active control system used in conjunction with a weak first-floor system, the actuators move the first floor of the building to the right when the earthquake tends to move it to the left and vice versa (Fig. 12.11). The building thus stands still from the first floor up while the foundation and the earth shake under it.

Due to the unavoidable time delays in electronic signal transmission and in the mechanical actions of the actuators, this ideal result cannot be achieved entirely; nevertheless, dramatic reductions have been demonstrated on shake-table models controlled by active antiseismic systems. One additional refined technique has thus entered the field of antiseismic design and we can look forward to the day when the catastrophes of the past may be almost entirely eliminated.

13

Social Consequences: Codes and Public Policy

It warn't no common shock, you bet!
Not one of your wavy kind,
But licket-y-split and rattle-ty-bang
Just which way it had a mind.

—W. H. CREIGHTON

The Story of Adam and Eve

The biblical story of Adam and Eve is, perhaps, the first unequivocal intimation of the fight between the will of man and the forces of nature. Tempted by the serpent (according to the psychoanalysts, their own subconscious), the first human couple eat the forbidden fruit of the tree of knowledge, become aware of their power to dominate the earth, and are compelled to abandon terrestrial paradise. What do they gain by their daring? Sacrificing their innocent happiness, they have gained, through the power of knowledge, the command of their destiny and the domination of nature.

After learning about the rebellion of our biblical ancestors, we had to wait over two thousand years of slow and demanding effort to gain the advantage, and learn the sorrows, of the biblical prophecy. Two geniuses, Galileo and Newton—the first by his trail-blaz-

ing experiments, the second by his astonishing abstractions—
opened the door of modern science and started its accelerating
race to the dominion of the earth. By our time, we have learned to
fight nature only too well and we have even developed the power
to destroy life on our planet. To avoid this final catastrophe, we
must learn to acquire a new sense of responsibility and temper our
insatiable craving for power. Yet, so far, we have barely started to
realize that our lack of humility is synonymous with collective
suicide.

These considerations explain, among others, our present quan-
dary concerning the consequences of our dramatically expanding
knowledge of seismology. For the first time in human history the
causes of these devastating physical phenomena, only a few centuries ago considered to be "acts of God", are now understood and it
is up to us to avoid their worst consequences. Unfortunately, our
recent hopes have not yet been entirely realized and confront us
with difficult financial, social and ethical dilemmas. On the one
hand, we are encouraged by having been able to forecast the location and magnitude of the last twelve strongest earthquakes on
earth; on the other hand, we are discouraged by our inability to
have predicted the time of their occurrence with significant accuracy. At present, uncertainties in the prediction of earthquakes
vary between ten and thirty years, and seismologists feel that
improving the reliability of their predictions may be very difficult.
But, in the meantime, earthquakes do not wait.

It's the Law!

As the codes of law are society's first line of defense against criminal behavior, building codes are the first ramparts against potential mistakes in the design, construction, maintenance and
restoration of buildings. Laws protecting building owners from
contractors go back to the harsh edicts of King Hammurabi of
Babylon (1792–1750 B.C.), carved for eternity on a marble column.
One of them states: "If the owner's son is killed because of faulty
construction, the son of the contractor shall be killed. If a slave of
the owner is killed, a slave of the contractor shall become the property of the owner."

Regulations to protect people from the hazards of earthquakes
have evolved slowly and usually in response to a catastrophe. In

Italy, the first seismic code was enacted in 1913 and was made mandatory only in those areas of the country where earthquakes had previously occurred. Los Angeles long ignored the hazards posed by earthquakes—at least until 1933: the Long Beach earthquake on March 10 of that year shook the city so badly that it led to the first requirements for the bracing of buildings. Because of the complete collapse of school buildings, it also led to statewide requirements to brace *all* public schools (that quake fortunately occurred after school hours, preventing a major loss of life among children).

Only after 1923 did the Japanese stringently impose earthquake-resistant building designs. Although devastated by the 1906 earthquake, San Francisco did not enact an antiseismic code until 1934. There have been few major earthquakes on the east coast of North America, and therefore little impetus to mandate seismic design provisions, except in the St. Lawrence basin around Montreal and the region around Charleston, South Carolina, both of which suffered destructive quakes in the nineteenth century. But the risk exists for a significant earthquake in the Northeast, so its major cities and states are now, slowly, enacting seismic provisions in building regulations. Code authorities have developed a zoned map of the United States defining the degree of seismic resistance required in different regions. The zones, ranging from 0, for no required seismic resistance, to 4 for the most stringent requirements, were established by considering historical precedent, risk and expected intensity of potential earthquakes. (Fig. 13.1).

Of course, modern building codes are not punitive: they seek to protect life first and then property. As such, they mandate specific requirements on loads, structures, materials and soils, but they leave to the law the penalties for nonadherence to their regulations. The basic justification behind the preventive rather than the punitive attitude of our building codes is dictated by the fact that, despite the efforts of code committees to write significant and unequivocal texts, codes are still open to technical interpretation and, hence, debate. Of course, codes also represent compromises between competing factions: builders who want the mildest regulations, public interest groups who want the most stringent regulations and politicians who waver in between. All recent codes are based on three principles: that buildings will resist only a) minor

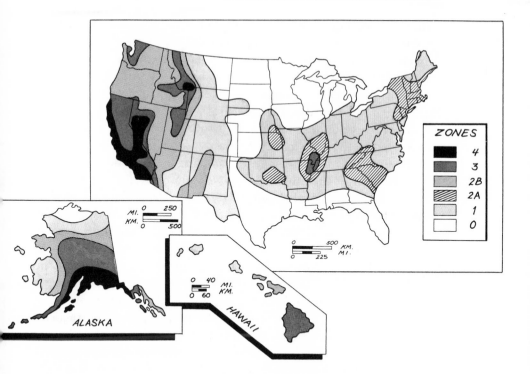

Fig. 13.1   Seismic Risk Map of the United States

earthquakes without damage; b) moderate earthquakes without
structural damage, but with some nonstructural damage; and c)
major earthquakes without collapse, but with both structural and
nonstructural damage.

In most countries of the world, one legal code is enacted by the
central government, but the U.S. federal government is constitu-
tionally prevented from imposing national requirements, leaving
to the independent states, counties and cities the responsibility of
enacting their own construction norms. The situation would be
chaotic but for the existence of a few codes that, because of their
thoroughness and significance, have acquired national acceptance.
Their requirements are introduced, with minor variations, into
most local codes. The stringent antiseismic provisions of the Uni-
form Building Code (UBC), developed in California, have been
widely accepted throughout the United States.

Two additional difficulties plague all codes. In order to be
adopted and actually implemented, they must be submitted to the

groups concerned with their new and generally more expensive requirements, and these groups require substantial amounts of time to study and grasp the significance of the changes. (Code committees often write commentaries to aid in the understanding of the revisions.) Even though codes encompass rapidly evolving fields, like structural dynamics and seismology, they cannot be modified too often lest they create confusion in the design professions. Hence, they are all, unavoidably, somewhat behind the scientific times. As if this were not enough, the introduction of the time variable in the study of the dynamic behavior of structures has greatly increased the complexity of their calculations and has required computer codes to be written that allow the average practitioner to solve such complex problems without running into the jungle of advanced mathematics. Finally, courses in seismology are outside most undergraduate engineering curricula. The resulting situation is, on the one hand, a most desirable and rapid increase in earthquake knowledge and, on the other, a practical impossibility for the structuralist to keep informed about theories he or she could and would use if only they were not changing so rapidly.

The reader should realize that *by law*, engineers are only expected to operate in accordance with "good engineering practice" and hence are not required to keep up with the latest seismological information, although, fortunately, most reputable engineers do. Moreover, despite the explicit caveats of all the building codes warning the users that the code requirements represent *minimum demands* for acceptable safety, seldom if ever are these demands exceeded in practice, mostly because of cost considerations. Engineers may be understandably afraid of losing clients if the cost of their designs is consistently higher than that of their competitors and they might well be reluctant to go beyond the code safety requirements, even when demanded by exceptional situations or by new knowledge that was not yet included in the codes. Engineering design, mysterious as it may appear to the layman, is a human activity, and humans come with varying levels of social consciousness and scientific knowledge. The minimal number of structural collapses occurring today in the technologically advanced countries and the successful resistance to the violent forces of nature of our highest high-rise buildings show both the successful influence of the seismic code provisions and the knowledge and morality of the construction practitioners.

What Can We Do?

Let us put ourselves in the shoes of the mayor of Los Angeles on the day in the not so distant future when the California Seismic Safety Commission announces: "We wish to inform you that there is an 80 percent chance of an earthquake of magnitude 7.8 to occur tomorrow in this area." Should the mayor immediately notify the population? Is this a psychologically correct move? Will the mobilization of the responsible police, fire and hospital departments be capable of avoiding the rush to safety of three million people and allow taking care of the injured? What of those who, for a variety of reasons, cannot be evacuated? If our forecast capacity of earthquake strength has improved rapidly and dramatically, will the antiseismic codes have caught up with it and mandated the retrofits needed to guarantee the safety of most, if not all, of the city structures? Would this have been economically feasible? Even if the forecast announced the quake in sufficient time and with adequate accuracy to organize the orderly evacuation of the city, would the alerted population behave civilly under the threat of death? Should the sick and the old or the children be evacuated first? Will the rich move out on their own before the poor?

Such unending questions, we believe, cannot be answered at the present time, or possibly ever, but the admirable discipline and coolness of the citizens of San Francisco during the 1989 Loma Prieta earthquake gives us hope that, with due warning and preparation, the well-known murderous stampedes of terrorized crowds would not necessarily take place. We now turn to the proposals enacted so far and the growing number of studies that aim to solve our newly solvable life-and-death problem.

California at Risk

In addition to what each one of us should do for ourselves and our families, certain measures to reduce earthquake risks can be and are initiated by the authorities of the federal government, the states, counties, towns and localities in areas of known seismic activity. The efficacy of such hazard-reduction measures increases with the population involved, but is most dependent on preventive, immediate *local* initiatives. Pervasive fatalistic attitudes, like those we previously quoted (see p. 124), constitute one of the major contributions to earthquake injuries and deaths.

At 5:54 P.M. on March 10, 1933, when most schoolchildren were home playing or perhaps even doing their homework, an earthquake of magnitude 6.3 struck the area around Long Beach, California. Violent tremors caused the collapse of several local schools and severe damage to all the others. Although the quake caused only 120 fatalities, the potential for a tragedy of monumental proportions had the quake struck only a few hours earlier, motivated the legislature to pass the Field Act, which set strict standards for building seismic resistance into new schools. When in 1971 San Fernando was struck by a quake of intensity similar to that of the Long Beach quake, a survey conducted after the event demonstrated the effectiveness of the statute: of five hundred new schools, not a single one was structurally damaged, while 10 percent of the schools designed before the Field Act were so severely damaged that they had to be demolished. (Our California readers will be relieved to know that *all* older schools have now been strengthened or rebuilt to correct seismic hazards.) Since first enacted, the Field Act has been extended to apply to private schools and other buildings serving the public. The San Fernando earthquake caused the collapse of the Olive View Hospital and spawned another governmental initiative, the evaluation and strengthening of all Veterans Administration hospitals throughout the United States.

Since 1986, California has implemented a comprehensive program to reduce earthquake hazards. The program, "California at Risk," is advisory in nature, but its authors have proposed seventy-two initiatives ranging from planning to education to future development and they have set priorities for hazard reduction. This has resulted in the implementation of legislation mandating the strengthening of unreinforced masonry (URM) buildings, the retrofitting of government buildings and the disclosure of seismic hazards in the sale of properties. The legislation also suggests that home water heaters be braced (see p. 171).

Both Los Angeles and San Francisco have mandated the retrofitting of all URM buildings[1] and, when needed, have provided

---

[1] Los Angeles permitted the construction of unreinforced masonry buildings after the 1906 San Francisco earthquake because they were believed to be fireproof. Such construction was abandoned after the 1933 Long Beach earthquake, because, by then, seismic hazards were better understood.

financial assistance either outright or through low-cost loans. Since such buildings constitute the core of these cities' oldest and most affordable rental units, the retrofitting program provides a social benefit as well as, most likely, the saving of lives. Of the 8,100 URM buildings in Los Angeles, 1,600 were residences with a total of 46,000 apartments: it is impressive to learn that by 1991, 55 percent of the buildings had complied with the ordinance, 13 percent were in progress, 12 percent had been demolished and only 20 percent had not been started.

"California at Risk" is evolving with the proposal of new legislation following each major seismic event: over a hundred earthquake-related bills were introduced after Loma Prieta in 1989, compared to only a handful two years earlier, and we expect a similar reaction following the 1994 Northridge earthquake. Even the 1988 earthquake in Armenia spurred California's legislators into action: they introduced four times as many seismic safety bills in 1989 as the year before. By replacing previously ad hoc measures, the California Earthquake Hazard Reduction Program provides the state's sixteen million residents who are at risk with information about whether their homes and workplaces will be safe from the ever-present danger of earthquakes.

Northridge

The residents of the northern Los Angeles suburb of Northridge, in the San Fernando Valley, were rudely awakened at 4:31 A.M. on January 17, 1994, by a magnitude 6.8 earthquake. Although "moderate" quakes of this magnitude occur in Southern California about every four years, the extraordinary damage caused by this thirty-second quake surprised even the experts: 11,000 residences were destroyed; an apartment building, a department store, a medical administration building and a two-year-old garage collapsed, and 1,300 other structures were too seriously damaged to be repaired; over 250 gas lines ruptured, igniting numerous intense fires; one of two aqueducts carrying water from the Sierra Nevada mountains ruptured; and 9 highway overpasses crumbled, totally disrupting traffic in this freeway-dependent region. It was the costliest natural disaster in U.S. history.

The extensive damage produced by the quake may be blamed on the nature of the fault, a "hidden" thrust fault (Fig. 13.2) causing

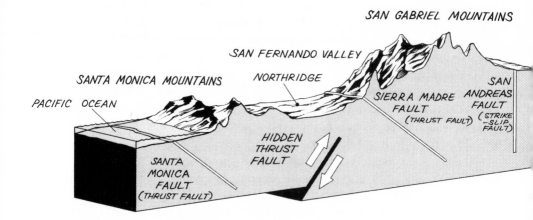

Fig. 13.2   Hidden Faults under Northridge

unusually large vertical accelerations (up to 1.8 times the acceleration due to gravity), and shaking a densely populated area in the
center of town right over the quake's epicenter. The quake exposed
a previously unknown, *almost horizontal* fault, that may be part of
a broad fault band, known as the Elysian Park System, which
extends across an arc north and east of Los Angeles (Fig. 13.3). As
the area was struck by over one thousand strong aftershocks, the
surrounding mountains rose by almost 300 mm (1 ft.) and seismologists expressed the opinion that future quakes of up to magnitude
7.5 might occur within the Elysian Park System. This prediction,
coupled with another worrisome fact—that there have been more
southern California earthquakes of magnitude above 6 since 1980
than at any time in this century—suggests that a "big one" may
soon occur in the Los Angeles area.

The Northridge earthquake provided some positive benefits by
testing those structures that had been upgraded following the 1971
San Fernando quake. Fewer than 20 percent of the state's bridges
had been retrofitted with reinforced columns and cable restraints
to prevent girders from sliding off the tops of the columns; reinforcement of the concrete freeway columns was still under way
when the quake struck. The columns already strengthened with
steel jackets came through virtually unscathed while 50 percent
of the unstrengthened columns were squashed by highway decks
dropping on them due to the large vertical accelerations. Similarly, many unretrofitted masonry buildings failed while those

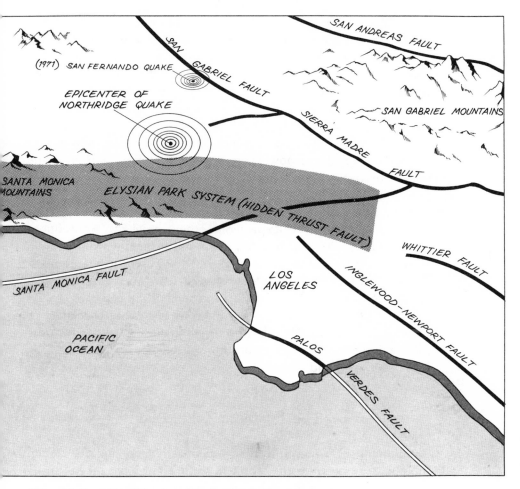

Fig. 13.3   The Elysian Park Fault System

that had been strengthened survived, to the great relief of their occupants.

The quake also served as the first live test for a base isolation installation at Los Angeles's University Hospital, an eight-story steel-framed structure. Horizontal earthquake forces equal to half the weight of the building, acting below its foundation, were attenuated by sixty-eight lead-and-rubber and eighty-one elastomeric isolators to 20 percent of the weight of the building, in the superstructure, confirming the predictions of both the computer models and shake-table tests. This 60 percent reduction in earthquake forces permitted the undamaged hospital to be the *only* fully operational facility available in the area right after the quake. In contrast, the Olive View Hospital, which had collapsed in the 1971 quake and had been rebuilt like a fortress, easily survived the quake, but its great stiffness caused the structure to shake in concert with the ground motions, damaging the sprinkler system and the hospital's contents, rendering it unusable for emergency services.

At the site of the Getty Center under construction on a hilltop above Los Angeles, ironworkers returning to their job after the quake discovered cracks in steel connections of two unfinished buildings. The connections, holding together a building's beams and columns, had been expected to stretch when subject to earthquake movements. Unexpectedly, they cracked. Subsequent inspection of 100 steel buildings ranging in height from 3 to 26 stories, in the most severely quake-affected area of Los Angeles, revealed that 75 percent suffered similar tears and cracks in some of their connections. None of the inspected buildings showed external signs of damage, raising serious concerns about the safety of the structures to withstand future earthquakes. After much study, a strengthening procedure for these weakened connections has been developed and is now (1995) being implemented.

The Japanese Tokai Plan

The Japanese, who could confront situations potentially much more damaging than the worst in California, have taken strong measures to cope with earthquake hazards.

The Tokai region (Fig. 13.4) covers the southern area of Japan between its two most populated cities, Tokyo and Osaka. The Tokai's southern coast faces the Suruga Trough, where the Philippine tectonic plate subducts the Eurasian plate (see p. 23). At intervals of about one hundred years this subduction generates a number of earthquakes, the last two of which occurred in 1944 and 1946 *without* relative slippage of the two plates. Because strain between the plates has been accumulating since the last earthquake in 1854, Japanese seismologists believe that a strong earthquake may be imminent in the Tokai region and they have advised the central government that an all-encompassing series of measures should be taken to reduce the damage of a potential seismic catastrophe. (The Japanese government, unlike that of the United States, has the authority to mandate specific measures to mitigate

Fig. 13.4   The Tokai Region

all kinds of catastrophes and to allocate the needed expenditures to the local governments, the prefectures [analogous to states] and the municipalities.) The first move in this direction was taken in 1969 with the establishment of the Coordinating Committee on Earthquake Prediction Research (CCEPR), which immediately intensified seismic investigations in the Tokai region. By 1979 the government had designated the region an *area of intensified measures*, covering 14 400 km² (5,400 sq. mi.) with about six million inhabitants in 170 cities, towns and villages. The committee assumed that the intensity of the expected earthquake will be greater than IX on the modified Mercalli scale,[2] roughly equivalent to a magnitude 7.9.

Following this designation, Japan's Metereological Agency (JMA) began monitoring seven on-shore stations and four ocean-bottom seismometers, electronically transmitting signals to the JMA headquarters in Tokyo; seventeen other locations were monitored by the National Research Center for Disaster Prevention and additional ones were monitored by local agencies, for a total of sixty-seven signals gathered from the Tokai and adjoining Kanto areas.

[2] Equivalent to greater than 6 on the Japanese scale, which encompasses values from 0 to 7.

The JMA has been charged with the responsibility of predicting the earthquake expected in the Tokai area and its director general has established an Earthquake Assessment Committee (EAC) to decide when the earthquake prediction should be communicated to the Japanese prime minister, who will announce it to the country. The members of the EAC are on twenty-four-hour notice to convene whenever the chief of the committee's observation section deems it necessary to obtain their decision. Whenever the EAC is called upon to convene, the members will be alerted by the electronic beepers they carry day and night, and they will need to be in Tokyo *within thirty minutes*. The decision to communicate an alert to the prime minister is the prerogative of the JMA chairman and does not require the consensus of the committee. As soon as the JMA committee is called to convene, the Japanese National Broadcasting Corporation will announce by radio and television to the Tokai area that the committee has been called into an emergency session.

A unanimous decision of the governmental cabinet is required before a warning can be issued by the prime minister. The entire prediction process is designed to take place in *less* than two hours and the prediction may be withdrawn if the earthquake does not occur within several days. In any case, the prediction of the earthquake will not be announced as a percent likelihood of its occurrence but will simply indicate that "there is a very strong likelihood of an earthquake occurring".

Dramatic measures will be taken following the prime minister's alert. Municipal governments will use sirens, fire alarms, loudspeaker trucks and radio and TV announcements to alert the population. Within two hours all road and railway traffic will be stopped. Telephone use will be limited and private calls will be automatically prevented. Schools will be evacuated. Banks will close, except to allow withdrawals from saving accounts.

Participation by the general population in the event of an alert is being fostered by educational programs in schools, newspaper special issues, radio and TV. By 1983 a voluntary program of community action, called *jishubo*, had registered 96 percent of the population in the Tokai area. They were given portable pumps, water purifiers, emergency floodlights, fire extinguishers, cooking pots, radio transmitters, storage tanks and other equipment. Lest the reader were to consider this high voluntary registration to *jishubo* an expression of extraordinary civil consciousness on the part of the Japanese, we must add that the number of active participants in *jishubo* activities is nowhere near 96 percent. Most of the registrants did not even become familiar with *jishubo* plans and with each passing month and year, popular commitment to preparedness wanes.

In order to appreciate the many reasons for this unusual national preparedness in the face of an expected seismic event, one must note that 85 percent of the housing in the Tokai area is of typical wooden construction, consisting of flexible posts and lintels with minor bracing, heavy roofs and brittle walls. Such construction is liable to collapse even under weak earthquakes: it has been estimated that of the 800,000 houses in the Tokai prefecture of Shizuoka alone, between 100,000 and 200,000 might be seriously damaged by the expected strong earthquake. Moreover, devastating fires would most probably destroy an estimated 275,000 of these buildings. Hence, the government's continuing support of the retrofit program to upgrade houses.

Finally, there is the serious danger of tsunamis all along the densely inhabited coast of the region. In the 1983 Nihonkai-chubu earthquake, 100 of the 104 deaths were due to a tsunami that took place two minutes after the earthquake. To minimize this danger, the coast is being protected by a tsunami seawall as high as 6 m (20 ft.), a single section of which extends 19 km (12 mi.), and by water gates preventing the flooding of inland areas. Analogous retrofit measures are being taken to minimize damage to buildings from liquefaction (see p. 197) and landslides, (see p. 114) as well as damage from any cause to the network of bridges of the entire area. All these measures are supplemented by a minutely organized evacuation plan. September 1 of each year is designated "Disaster Prevention Day": exercises on that date are central to the effort to achieve coordination among the many jurisdictions. A return to normalcy after a strong earthquake is not expected for substantial periods of time, but essential services are guaranteed to become available from two days to a week after the earthquake occurrence.

The tragic record of earthquake damage in Japan makes obvious the country's need for extraordinary plans like the Tokai prediction and prevention plan. Yet U.S. readers should not feel complacent about the earthquake situation in California or other seismically active areas of the United States, and should be aware of the important, even if more limited, plans enacted in our country. The American Committee on the Anticipated Tokai Earthquake concluded in a 1984 report published after a visit to the Tokai region, that "there is a striking parallel between the situation in the Tokai region and the situation in the Los Angeles basin, where a long-term prediction has been made of a great earthquake on the south-central segment of the San Andreas fault. The Japanese are now coping with a situation that Southern California may have to cope with in the near future",[3] particularly after the unexpected Yucca Valley earthquakes of 1992 and the discovery of new or extended faults in that area (see p. 123).

The Great Hanshin Earthquake

The residents of Kobe, a port city 430 km (275 mi.) south of Tokyo, were violently awakened without warning, at 5:46 A.M. on Tuesday, January 17, 1995, by the most powerful tremor to strike an urban area of Japan since the disastrous 1923 Tokyo earthquake (see Chapter 7). The quake of magnitude 7.2 occurred, not in the Tokai region as expected, but further south in the Kansai region that includes Japan's third largest city, Osaka and the ancient capital of Kyoto. The focus of the quake lay less than 20 km (12 mi.) below Awaji-shima, an island in the inland sea facing Kobe and Osaka. As the residents of Northridge learned a year earlier, such a shallow earthquake is very destructive and causes extensive damage, although its effects diminish rapidly with increasing distance from the epicenter. The city of Osaka, only 32 km (20 mi.) east of Kobe, suffered only minor damage as did the new Kansai International airport built on an artificial island in Osaka Bay.

In the city of Kobe, the twenty-second duration shock had disastrous consequences. It toppled the main elevated highway linking Kobe to Osaka in five places with one 600 m (2000 ft.) long section leaning on its side like a chain of fallen mushrooms and with its central concrete columns shattered revealing too few tie bars to hold the vertical bars together. It destroyed or damaged more than 46,000 buildings, many hastily built after the Second World War without adequate antiseismic strength. It cut gas,

---

[3] *Proceedings of the U.S.-Japan Workshop on Urban Earthquake Hazards Reduction*, Publication no. 85-03, July 1985, Earthquake Engineering Research Institute.

water and electric service, derailed seven trains, damaged the line of the high speed "bullet" train (had the train been running, it would have been automatically stopped by signals sent from seismographs). The destruction left 300,000 of the city's 1.5 million people homeless.

Of 186 berths in the port of Kobe, only 8 were operational after the quake, reducing the entire cargo handling capacity of Japan by over 10 percent. Broken gas pipes throughout Kobe caused fires to erupt and, as water was cut off in many parts of the city, many of these fires burned until all consumable material was reduced to ashes. Newer structures were deformed by the quake although they did not collapse, proving once again that earthquake-resistant structures are not earthquake proof. This quake, as many previous ones, will result in a re-evaluation and revision of seismic codes in Japan and other countries.

The human toll was second only to that of the great Tokyo quake with over 5,000 dead and over 20,000 injured. In the wake of continuing aftershocks, one terrified resident stated, "I haven't gone through an experience like this since the Second World War when the city was virtually destroyed by bombing". The severity of the physical damage and the lack of any warning surprised both Japanese seismologists and engineers who, after viewing the fallen freeways and crushed buildings in Northridge, California, the previous year, had proudly declared that such failures could not happen in modern Japan.

The residents of Kobe were spared the devastation from a tsunami that often follows such a quake by two lucky conditions: The quake was the result of a 1.5 m (5 ft.) *horizontal* strike-slip dispacement extending about 9 km (5.6 mi.) along the Nojima fault on Awaji island, while a tsunami is usually caused by a *vertical* displacement of the earth's crust, and the relatively shallow water surrounding the island that limited the wave height of a potential tsunami.

In nearby Kyoto both the fourteenth-century Golden Pavilion and the tenth-century Pagoda of the Daigoji temple cracked as a result of the quake. In addition, a number of statues of the Buddah were toppled or otherwise slightly damaged. The relatively minor damage to these historic structures was the only bright sign in an otherwise horrifying landscape due to this, *the most costly ever* natural disaster.

Cry Wolf

Predictions of an earthquake that turn out to be false can cause trauma to local populations and seriously damage the credibility of seismologists. When A. Inamura, a young Japanese seismologist, presented his predictions for the Kanto earthquake of 1923 (see p. 82), he was not believed because of a fear of social unrest in the country and a lack of confidence among more experienced seismologists in the scientific basis of his forecast, based on the novel *seismic gap* (period of return) concept. Of course, he was not able to pinpoint the timing of the expected quake within a generation. How long can a population remain in a state of readyness?

The seismologists of the Chinese State Seismological Bureau were confronted with just such a situation in 1975. For several years the seismic activity around Haicheng had increased dramatically, causing the seismologists to warn of a major quake that, based on the pattern of observed earthquakes, *should* occur in the first half of 1975. Suddenly, in the late morning of February 4, all seismic activity stopped and was interpreted by the seismologists as a warning of an imminent event that had to be communicated to the general population . . . but what if they were wrong? Fortunately, they bravely decided to act and issued a warning at 2 P.M., just five hours before a major shock struck the Haicheng area. Their action saved innumerable lives: 90 percent of all the homes in the area (of unreinforced masonry construction), collapsed while their occupants, bundled against the winter winds, were safely camped outdoors, in response to the forecast announcement.

On the other side of the coin, in August 1976, a month after the Tangshan earthquake, a warning was issued by the Chinese authorities to the residents of Kwangtung Province about an impending earthquake. For two months people slept in tents and spread their fear to the inhabitants of nearby Hong Kong. But nothing happened!

In 1980, Drs. Brian Brady and William Spence, two American scientists working for the U.S. Geological Survey, announced at a scientific meeting in Argentina that three catastrophic quakes would strike the coast of Peru around June 28, August 10 and September 16 of the following year. These extraordinarily precise predictions presented by American government scientists caused widespread alarm throughout Peru. Even though the predictions were not endorsed by a government agency, the general population believed them—the predictions had been made by "distinguished persons." The Peruvian government demanded a review of the predictions by the U.S. National Earthquake Prediction Council, which, in January 1981, repudiated them, citing no substantiating data. Nevertheless, the damage had been done and the "Brady-Spence predictions" severely damaged respect for the scientific establishment. So it goes in seismology. Predictions by amateurs, fanatics and even some too-enthusiastic seismologists can be written off, but greater control of scientific predictions, without impinging on scientific freedom, is necessary to avoid social apathy from "cry wolf" situations.

14

Risk and Preparedness

A bad earthquake at once destroys our oldest asociations. The earth, the very emblem of solidity, has moved beneath our feet like a fluid; one second of time has created in the mind a strange idea of insecurity which hours of reflection would not have produced.

—CHARLES DARWIN

While Waiting for the Big One!

We now turn to a brief series of answers to the most common questions asked about earthquakes. These cannot substitute for thorough and useful information offered in books of earthquake advice to the layperson and in the numerous pamphlets on the subject freely distributed, among others, by the California authorities. We present this succinct list of answers to such questions only to start the reader on the path of earthquake education.

1. *"We do not live in California. Do we have to worry about earthquakes?"*

You do. Besides the states bordering California, other American states have been subjected to earthquakes. In fact, the strongest and most devastating earthquake ever in the United States occurred near St. Louis, Missouri, in 1803, and the largest quake on the East Coast occurred in Charleston, South Carolina, in 1886. Some parts of the United States are immune from earthquakes, others are not (see p. 149). California suffers from the most frequent and, usually, strongest earthquakes.

2. *"What are the latest predictions for the time of arrival and the magnitude of the 'Big One' in California?"*

The prediction at the time of this writing (1994) is that the "Big One" has a 67 percent chance of occuring any time within thirty years and that its magnitude *may be* above 7. Contrary to previous predictions, it might occur along the southern portion of the San Andreas Fault, as shown by the 1991 Yucca Valley earthquakes (see p. 123).

3. *"We live in a one-family wood-frame house in a seismic area. Are we in great danger?"*

Yes and no. Wood is an excellent earthquake material if it is not attacked by rot and if the joints between horizontal and vertical structural components are correctly executed. If there is a chance of rot in the wood or if the nails show signs of rust, the house should be inspected by an engineer. If the house was designed according to the latest governing code and such problems are not present, you are probably as well off as in any other seismically designed building, provided the house is well anchored to its foundation. If built before 1970, you may have to have your house bolted down and the "cripple walls", between the footing and the floor, braced (Fig. 14.1).

4. *"We live on the upper floor of a ten-story reinforced concrete building. Is it safe against earthquakes?"*

If the building was designed after the 1950s, that is, in accordance with modern antiseismic codes, you are probably as safe as anybody can be from structural collapse. But no one can guarantee that some parts of the building may not break off and injure you.

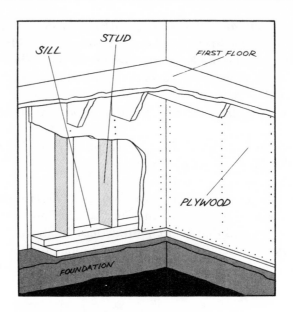

Fig. 14.1   Reinforcing Cripple Walls

5. *"We live in New York City. Are we in any danger from earth-quakes?"*

Yes, but there is a low chance of occurrence of an earthquake and its expected magnitude is below 7, so that structural collapse is unlikely in the city or in any part of *southern* New York State. An antiseismic code was signed into law by the mayor of New York City in 1995.

6. *"We live in upstate New York. Is there any danger of earthquakes here?"*

Yes. The St. Lawrence River region is seismically active and has experienced magnitude 7 earthquakes, with a risk of future recurrence, which led to the requirement that some public buildings, like hospitals and government buildings, be retrofitted.

7. *"My office is on a high floor of a skyscraper in San Francisco. Am I in danger?"*

If the big one occurs while you are still alive and if it is of magnitude 8 or above, structural damage may occur to your high-rise building. But San Francisco high-rises behaved admirably in the 1989 Loma Prieta earthquake that was of magnitude 7.1.

8. *"Our summer home is built on the side of a steep hill on the Pacific Coast. Are we safe from earthquakes?"*

Earthquakes often induce landslides in unstable, steep slopes, such as those found in the hills around Los Angeles and along the Pacific Coast Highway, and such areas should obviously be excluded from construction. If the soil under your house is loose, and particularly if wet, it may slide during an earthquake whereas rock will not. If you are in doubt, ask a soils engineer to determine the type of soil and the kind of foundation of your house.

9. *"We live in a masonry landmark building, built in 1890. We are told that it is located in a zone 3 area. [see Fig. 13.1.] Should we worry?"*

You might want to worry, but you should also realize that retrofitting a landmark building is complicated, expensive and particularly difficult because of the requirements of landmark commissions. We suggest you have the building checked both for your safety and for the historical value of the building. Buildings of historic value, such as the 1893 Salt Lake City and County Building, have been retrofitted with seismic dampers (see p. 134).

10. *"I have been transferred by my company to a job in Turkey. Do I have to worry about earthquakes?"*

Yes, strong earthquakes have occurred in Turkey (see table 6.1). Find out whether your office and residence satisfy the local, modern seismic codes.

11. *"Why do such large numbers of people die in earthquakes in the underdeveloped countries?"*

Because many underdeveloped countries have large rural populations. People usually live in precariously built huts, often wooden huts with heavy tile or thatched roofs. They are killed most often by the collapse of the roofs in strong earthquakes. Moreover, the urban populations in many underdeveloped countries are among the largest in the world and most of the city dwellers live in poorly built houses that crumble even under relatively weak earthquakes (and may easily burn).

12. *"I live in California. Where can I obtain advice on earthquakes?"*

The Earthquake Section of the U.S. Geological Survey, 345 Middlefield Road, Menlo Park, CA 94025, is the best source of general information. Some of their materials are available in English, Spanish, Chinese, Braille and on recordings for the blind. Local authorities in California are also excellent sources of earthquake information.

13. *"We live in a ten-story masonry building in a zone 3 area. [see Fig. 13.1.] Is it dangerous?"*

If the building is entirely built with unreinforced masonry, you live in one of the worst structures you could inhabit in an earthquake area. But, if the building has a steel or a reinforced concrete frame and *all* its walls, *including the ground-floor walls,* are of reinforced masonry, it may be safer than usually assumed, because a correctly designed frame should prevent the building's collapse, while the masonry walls would probably crack in a strong earthquake and, in so doing, absorb by friction a substantial part of the earthquake energy.

14. *"I often attend sports events in a stadium with a fabric-type roof. Am I in danger during an earthquake or a hurricane?"*

Fabric-type roofs are not particularly dangerous in themselves in strong earthquakes because they are very light and, hence, have a small mass. But the catwalks, heavy lights and loudspeakers hanging from them obviously present a danger. A hurricane may well tear the fabric off a roof, but if a fabric roof is air-supported and only punctured, it will deflate slowly and provide ample time for evacuation.

15. *"Should I run out of my office building during an earthquake?"*

Never, because, unless the entire building collapses (an unlikely event), you run a much greater risk of getting hurt if you try to flee. Hide under a strong table or a desk, grab a table leg or anything well anchored to a wall and *stay put* (Fig. 14.2). A large number of casualties occur to the ill-advised and the scared who try to run out of a swaying building and either trip on an unexpected object, fall down the stairs, get hit by a non-structural element falling from the building's facade once on a sidewalk or are run down by a panicky driver once they reach the street.

16. *"I live in a zone 4 area [see Fig. 13.1]. Are there any risk-prevention measures I should take against earthquakes?"*

Get information from the earthquake authorities in your area, but meanwhile do the following *right now.*

- Anchor to the floor or to the studs in the walls *(not the sheet-rock partitions)* any heavy furniture or pieces of equipment (heavy chests of drawers, gas water heaters [Fig. 14.3], computers).
- Fasten pictures and other heavy objects hanging on the walls.
- Latch doors and cabinet doors.
- Fasten and / or restrain valuable and heavy equipment (computers, china) to secured shelves or anchored tables.
- Store emergency supplies: first-aid kit, water, flashlights and batteries, battery-operated radio and fire extinguisher.
- Rehearse the earthquake plan you will follow when the earthquake begins.
- Have your home or building checked for adherence to the governing codes.
- Check with school authorities to ensure they have taken similar measures.

17. *"What do I do if I'm in a car when the earthquake strikes?"*

Move to the shoulder of the road and, if possible, stop the car away from power lines, bridges and buildings. Stay in your car until the shaking stops.

Fig. 14.2   Safe Positions in an Earthquake

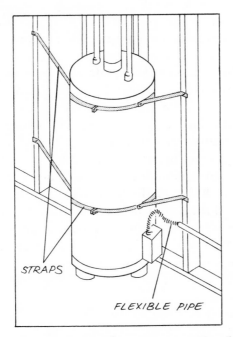

Fig. 14.3   Restraining a Tank against an Earthquake

18. *"I live in Utah. Am I in danger of earthquakes?"*

Yes, if you are one of the 85 percent of Utah residents who live along the Wasatch Fault region. A big, magnitude 7.5 quake occurs along some segment of the fault every 350 to 450 years and the last one occured about 400 years ago. Somewhat smaller, magnitude 6.5 quakes take place there every seven years.

19. *"I am buying a piece of property in California and received a form saying that it is not in a 'Special Study Zone.' Does that mean no earthquake hazard?"*

No. Special Study Zones refer only to the potential that the property lies on or near a rupture, but does not deal with other geologic hazards such as liquefaction, landslide or ground shaking, all of which are very dangerous.

20. *"What are the causes of injury in an earthquake?"*

A table compiled after the 1989 San Francisco earthquake lists the frequency of disabling injuries (that is, those needing medical attention *and* hospital treatment), and gives new information on this matter.[1] The table indicates that the largest number of disabling injuries (28.9 percent) was due to falls, the second largest (17.5 percent) to being thrown or bumped into an object, the third largest (13.2 percent) to being hit by a falling object and the fourth largest (11.1 percent) to hitting or bumping into an object. But the table also indicates that *all* injuries due to structural collapse add up to only 4.6 percent, even after inclusion of the 2.5 percent due to the collapse of one section of the double-deck Nimitz Freeway in Oakland (see p. 95). Thus, contrary to common wisdom, in the Loma Prieta earthquake nonfatal injuries due to nonstructural causes were twenty-one times more numerous that those due to structural causes.

21. *"What happens to pipelines or structures built across an active fault?"*

A photograph of the 1906 San Francisco earthquake shows a road displaced laterally as it crossed the fault. If a building had been built at that location, it would have been sheared apart (Fig. 14.4). It is obviously not healthy to build across a fault! Few structures are actually built over faults, as evidenced by the statistic that only 1 percent of all seismic-caused losses in California in the last thirty years were attributed to fault displacement. Nevertheless, in 1972, the state of California passed the Alquist-Priola Act, which restricted construction in a 180–390 m (600–1,300 ft.) wide strip of land centered on active fault traces.

22. *"I plan to build a new factory near the shoreline on what looks to me like sand. What should I do to prevent the building from sinking?"*

We have emphasized the particular danger of liquefaction during an earthquake in sandy soils permeated by water and in landfills (see p. 107), which should discourage construction on

---

[1] Improving Measures to Reduce Earthquake Casualties," Michael E. Durkin and Charles C. Thiel, Jr. *Earthquake Spectra* (Feb. 1992): pp. 95–113.

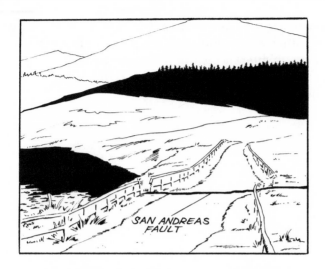

Fig. 14.4   Offset of Road Caused by Fault Slip

such sites. But, buildings on valuable property in the heart of some cities, like San Francisco and Mexico City, can be made safe using long piles driven down to a solid-bearing layer.

23. *"I live in Seattle and have never seen or heard of an earthquake fault here but I am told that I should be concerned about the possibility of a quake. Why?"*

The San Andreas Fault has clearly mapped its existence in the form of a visible rupture, but not all faults are as accommodating: those in Japan and Chile, two of the world's most active fault regions, lack identifiable surface traces. Offshore faults and those that are deeply buried under alluvial deposits, such as around Charleston, South Carolina, are not visible. The New Madrid Fault in the central United States is covered by a thick layer of sediment and the northwestern United States, where you live, hides its faults beneath a thick blanket of volcanic ash.

24. *"I'm a New York State legislator and am aware of the seismic hazard in my state. But, I don't remember ever experiencing an earthquake here. What should I do to protect my constituents?"*

Although some sections of New York State are vulnerable to serious earthquakes, they may never happen in your lifetime.

Nevertheless, you can propose legislation strengthening structures sufficiently to mitigate injury or avoid death of your constituents, while accepting a greater level of damage to
structures than is acceptable in California, where serious
earthquakes occur more frequently.

15

Echoes of the Big Bang

This is the way the world ends
Not with a bang but a whimper.

—T. S. Eliot

The Slow Birth of Science

Homo sapiens,[1] man the knower, wondered at the mystery of creation ever since he appeared on earth forty thousand years ago, only fourteen hundred generations before our own. His universe was then limited to the perception of the earth and the sky, yet the oldest myths of mankind provide strong evidence of his desire to explain this haunting mystery. Echoing much older beliefs, the Babylonian cuneiform tablets of five thousand years ago tell us in the epic of the *Enuma Elish* about the fights between the young gods of order, led by Marduk (the creator of mankind, light and life), and the old gods of chaos, that resulted in the victory of the young gods and the creation of a meaningful world. Three thou-

[1] Officially *Homo sapiens sapiens,* to differentiate today's man from *Homo sapiens neanderthalensis,* who appeared on earth about 100,000 years ago.

sand years ago the Egyptians believed that the universe was a flat disk, with the plain of the Nile at its center, supported by a circle of distant mountains. Protected by a flat sky, their universe had been created out of primeval chaos by the ram god Khnum (or, perhaps, by the pure spirit Ptah or, maybe, the sun god Ra and his descendants, the air god Shu, the sky god Nut and the earth god Geb). And only twenty-nine centuries ago the book of Genesis in the Jewish Bible described in remarkable detail how the one ineffable God created in six days the universe in all of its parts and, pleased with His creation, rested on the seventh, the Sabbath.

But we, children of the modern world, have a single myth, that of science, the searcher of mysteries, and look back in amazement

Fig. 15.1   Galileo's Gravity Experiment

to the early philosopher / scientists and to their understanding of the physical principles governing our world.

In about 350 B.C., Aristotle believed that bodies were naturally at rest and that a force had to be applied to move them. He concluded that, if you dropped two objects, the heavier would fall faster than the lighter because the power pulling them down (what we call *gravity*, the force pulling us toward the center of the earth) was larger for the heavier object. Aristotle the philosopher never thought of testing his hypothesis experimentally. But in 1590, when Galileo Galilei (1564–1642), the Italian astronomer, mathematician and physicist, doubted Aristotle's assumption and did drop identical balls of different materials and hence, different weights, from the top of the leaning tower of Pisa, he observed that all the balls fell at the same rate (Fig. 15.1). By breaking with established beliefs and disproving by experiment Aristotle's theory, Galileo became unpopular with his colleagues at the University of Pisa and had to resign his professorship. But, to his glory, he had discovered the first principle of *dynamics*, the study of the motion of matter.

Isaac Newton (1642–1727), after supposedly being awakened by an apple falling on his head and actually reviewing Galileo's measurements, in 1687 proposed, to his greater glory, three hypotheses or laws that govern the motion of bodies and which opened the path to modern science.

Newton's laws explained the motion of any body interacting with other bodies, be it the fall of an apple or the motion of heavenly bodies through the universe. So powerful were his principles that it was not until 1905, when Albert Einstein (1879–1955) proposed the special theory of relativity (of which Newton's theory is a particular case), that a more accurate explanation of the relation between force, mass and motion opened a wider door to both the infinitesimal world of atomic physics and the unbounded world of the universe. In what was to become the key to unlocking the mystery of the origin of the universe, Einstein concluded that nothing can travel with a speed faster than that of light *(c)*, and that mass *(M)* and energy *(E)* are equivalent forms of the same physical entity, as quantified by his famous equation $E = Mc^2$. Between 1906 and 1916, Einstein postulated additional hypotheses (**Einstein's hypotheses**) in his general theory of relativity and suggested considering the universe a four-dimensional space, a suggestion that revolutionized the field of physics.

Quantum theory, first proposed by Max Planck (1858–1947) in 1900 and later developed by a number of other scientists, took a *probabilistic approach* to physics and succeeded in finding answers to some of the most puzzling questions of creation. And this is where we are today in trying to understand our greatest mystery, the creation of the universe, although a number of new "string theories" are being proposed.

The Cosmic Egg

Contemporary physicists now believe that the universe started as an infinitely small, unimaginably dense and hot mass, containing all existing matter and energy. Incredible as it may seem, this nucleus of the universe occupied a space smaller than a single atom (with a diameter less than one thousandth of a billionth of a billionth of a millimeter), and within it the **four fundamental forces** governing today's physical world (gravitation, electromagnetism and the weak and the strong nuclear forces) were one.

For some as yet unexplained reason, the nucleus of the universe suddenly began to expand and swelled so rapidly that it threw out particles in all directions in an explosion the Russian-American physicist George Gamow (1904–1968) named the *Big Bang.* Because scientists have not yet been able to include an explanation of the force of gravity in the mathematical equations explaining the other three universal forces (Einstein himself tried this in vain for thirty years), we cannot apply today's physics from the very instant of the Big Bang: we must wait an inconceivably short amount of time, called the **Planck time,** for our physics to become valid. But, we know how to describe the behavior of the universe from *that* time to the present and even into the future although, so far, not forever. Due to uncertainties in the measure of the ratio of two physical quantities (deuterium and helium) we shall describe later, a serious question arises about the end of the universe: It may keep expanding forever, it may stop expanding or it may even shrink back to a point in a "Big Crunch" and explode again in an infinite series of Bangs and Crunches. Astrophysicists hope to make a decision about the destiny of our universe in the near future.

At this point, an inquisitive reader may ask, "I know that according to Einstein's general theory of relativity, light is bent by the gravitational field of a mass, as proven in 1919 when the rays of a star were seen to bend toward the sun during a total eclipse (Fig. 15.2). I understand that quantum theory states that if I know exactly where I am, I cannot know exactly how I am moving, and if I know exactly how I am moving, I cannot know exactly where I am. I even understand that, according to quantum theory, a cat can be inside and outside a box at the same time, although I don't quite fathom this amazing statement. But can you tell me how we found out what happened fifteen to twenty billion years ago, when you say the universe was created in the Big Bang?" Fortunately, we can answer this question, because at least four pieces of good experimental evidence confirm the correctness of the Big Bang theory.

The first piece of evidence relies on the inventions of two of the greatest scientists of all time, Galileo and Newton. Galileo, by using a combination of lenses originally devised by Dutch oculists, first built the telescope and saw the rings of Saturn; and Newton, by first using triangular glass prisms, brought the rainbow into the laboratory, showing that sunlight consists of a **spectrum** of colors (Fig. 15.3). From these modest beginnings, modern radio telescopes, in orbit around the earth, have allowed us to catch radiation with wavelengths below the visible red (infrared) and above the visible blue (ultraviolet) and to notice that there is a shift toward the red direction in the **light spectrum** of the stars in the **galaxies**[2] with respect to that of light on earth (Fig. 15.4). This phenomenon is analogous to the **Doppler effect** in which, for example, the pitch of a locomotive whistle increases as the locomotive moves toward us and decreases as the locomotive moves away from us. Edwin P. Hubble (1889–1953), the American lawyer turned astronomer, observed such a shift toward the red wavelengths in the light of the stars in distant galaxies and was the first to conclude that these galaxies were moving away from us. He also discovered that the more distant the galaxies were from us, the greater the **red shift,** and therefore, the faster their motion away from us. In explaining Hubble's observations, the Belgian priest and astronomer Georges Lemaître (1894–1966), in 1929 proposed that, since the galaxies were all moving outward in all directions, at some time in the past the entire universe must have suddenly started expanding from one spot he called the "cosmic egg". This sudden expansion is what we now call the Big Bang.

[2] From the Greek *gala* for milk, hence our Milky Way.

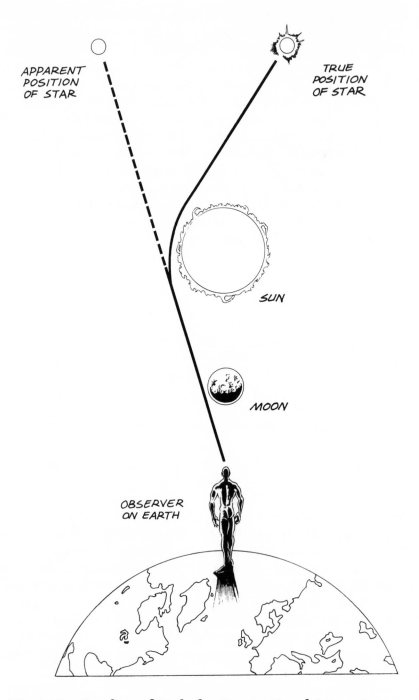

APPARENT
POSITION
OF STAR

TRUE
POSITION
OF STAR

SUN

MOON

OBSERVER
ON EARTH

Fig. 15.2   Bending of Light by Gravitational Attraction

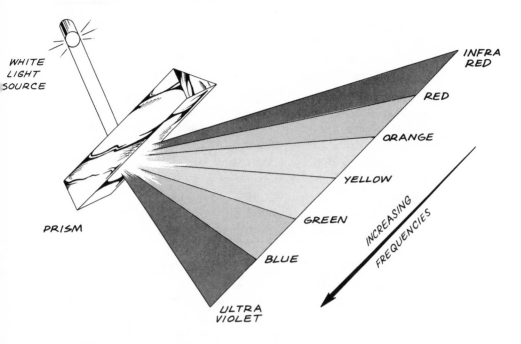

Fig. 15.3   Light Spectrum

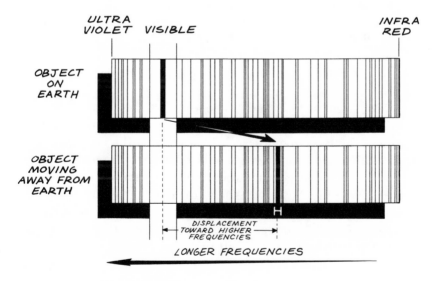

Fig. 15.4   Spectral Shift

The second piece of physical evidence of the Big Bang's correctness is due to a serendipitous discovery. Since 1948, George Gamow and his coworkers had surmised that, as the temperature of the universe kept decreasing after the Big Bang, and since the wavelength of radiation depends on temperature, at its present measured temperature, radiation of a certain wavelength should fill the universe uniformly. In 1965, while trying to measure the wavelength radiation of distant stars with their radio telescope, Arno Penzias and Robert Wilson of the Bell Telephone Laboratories caught a slight "radiation noise" that came equally from all directions and, sure enough, it was the remnant of the Big Bang radiation predicted by Gamow! Though most scientists were satisfied with this evidence of the Big Bang, two further pieces of evidence have recently confirmed it.

Nuclear physics shows that among the chemical elements resulting from the Big Bang, hydrogen, the lightest of all atoms and the most common in the universe, has an **isotope** called *deuterium* that, at the present stage of evolution of the universe, should be found in a *given ratio* to the amount of helium, the lightest element after hydrogen. Astrophysicists have recently determined that the experimentally measured ratio of deuterium to helium in the universe is *exactly* equal to the ratio predicted by the Big Bang Theory.

Finally, it has been found that the value of the temperature of the universe determined experimentally is exactly equal to that predicted by the Big Bang Theory. Could one demand more? Unsatisfied astrophysicists all over the world are now trying to solve one more mystery: why the clumps of matter in protogalaxies[3] were formed by gravity and why they are distributed the way they are in the universe.

Most scientists are ready to believe that these and other, more subtle pieces of evidence amply confirm the Hubble-Lemaître-Gamow "Big Bang hypothesis" . . . at least until a better theory of the creation of the universe is offered and confirmed by experiment. Let us not forget that science does not *explain* physical phenomena: it only describes them by making assumptions that physicists try to confirm through experiments. We believe that no scientist would be happier than Newton to know that his "assumptions" (we often mistakenly call them "laws") had been improved by Einstein, although he may be smiling benignly in noticing that,

[3] From the Greek *proto*, for first.

so far, his gravitation has refused to get mixed up with the other three universal forces. (We have a hunch he may also be thinking: "Sometime soon my gravity will probably join them".)

Let us now find out how the expansion of the universe led to the creation of the stars, including our own sun, and that of the planets,[4] including our own earth, and why our still evolving earth is never quiet, and occasionally rattles and shakes in earthquakes and blows off steam in volcanic eruptions. It makes a fascinating story and explains all the stories you have read in the pages of this book.

The Awesome Size of the Universe

After the Big Bang explosion, the universe began to expand, throwing particles of matter into space in all directions at such an unimaginable speed and to such incredible distances that the shortest unit of time (the second) and the longest unit of length (the kilometer or the mile) are unable to express the immensity of this mind-boggling physical process. New units had to be devised to describe time and distance in terms more appropriate to the immensity of the universe.

To do so, we could first extend the list of words recently made familiar by our economists in discussing national budgets. We had the thousands and the millions, and now have the billions and the trillions; so why not quadrillions and quintillions or even tentillions and fiftillions? The rub is that, luckily, these new words have not yet been needed "financially", and hence, have not become familiar.

To devise the needed units we could also adopt the shorthand symbols suggested by the philosopher / mathematician René Descartes (1596–1650), who, fed up with having to write "12×12" ten times in a row in a letter to a friend, suggested that 12^{10} would be a simpler way of putting on paper this insufferably long multiplication.[5] Thus, 10×10 fifty times, that is, 1 followed by fifty zeros, would become simply 10^{50}. Using the same shorthand to represent a minuscule time or length, instead of writing "one millionth",

---

[4] From the Greek *planētes*, for "wanderer."

[5] He originally suggested using a symbol with a subscript, 12_{10}, but it was soon changed to the current form with a superscript, 12^{10}.

which is one divided by "1 followed by six zeros", or "a fiftillionth", one divided by "1 followed by 50 zeros", mathematicians chose to write 10^{-6} and 10^{-50}. All scientists have adopted this practical form of notation, known as *exponential notation,* and we will also use it in describing the infinitely large dimensions and the infinitely short times encountered in the universe.

Finally, astronomers and astrophysicists needed entirely new units of length to *reasonably* express the size of the universe and, by multiplying the velocity of light, $c = 299\ 792$ km / sec (186,282 mi. / sec.) by the distance light travels in a year, that is, in 31,526,000 seconds, they obtained a unit of length called a *light-year* (abbreviated "lt-yr"), that fits their needs and is approximately 9.5 trillion km (6 trillion mi.). (Eventually, they adopted an even larger unit of length, the *parsec,* equal to 3.26 light years).

Armed with these useful notations and units, all of us are now better able to understand how the universe evolved after the Big Bang explosion.

An Instant Later

One hundred million Planck times (10^{-35} sec) after the Big Bang, the universe had already expanded to 10^{50} times its original volume. As it expanded, it cooled into a "soup", allowing the nuclear components of the "potential" atoms, the **protons** and **neutrons,** to combine together into real atoms and real energy and then form the chemical elements that make up all matter.

Within one minute after the Big Bang the nuclei of hydrogen atoms (the positively charged mass in atoms) in the "soup" had already begun to "fuse" together, giving birth to helium and energy in nuclear fusion reactions. In these reactions, two nuclei are combined into one nucleus of less weight than the sum of the weights of the two nuclei and a large amount of energy is released, according to Einstein's equation $E = mc^2$ (the same reactions give hydrogen bombs their destructive capacity).

After only one hundred years, more complex atoms were generated by hydrogen and its **isotopes;** matter and energy decoupled into separate aspects of the same physical entity and the universe became transparent! Its temperature cooled from the original $10^{27\circ}$ **Kelvin** to a modest 6,000° K.

A million or more years after the Big Bang, as the universe continued expanding and cooling, gases and eventually solids and liquids were formed. This appeared to take place uniformly in the universe but, just as a cloud of dust appears evenly dense to the naked eye only because one cannot count the different number of dust particles in its different regions, some regions of the universe gathered more particles and were denser than others. The newly created force of gravity acting on these particles slowly attracted the less dense gases around the denser spots and condensed them into disks called *protogalaxies,* from which the future galaxies would evolve (the Hubble Space Telescope has detected the intergalactic medium, the gas that collapsed into lumps forming the early galaxies).

Because of their increased mass and consequently increased gravitational force, these disks of matter slowed their outward expansion and began to collapse inward. Sliding past each other, the disks started rotating as the gravitational force interacted between them. This rotation prevented each disk of matter from completely collapsing toward its center because, as each of them became more tightly bound and smaller, their spinning increased, just as ice skaters spin faster as they pull their arms closer to the body. Eventually a state of balance was achieved between the force of gravity attracting together all the specks of matter within a disk and the *centrifugal*[6] force that tried to pull them out (Fig. 15.5).

In this way, about two billion years after the Big Bang, rotating lens-shaped galaxies, containing billions of stars, were formed. Among them was the Milky Way, of which our own solar system is a member.

When we look up to the silent dome of the sky on a clear night, millions, indeed billions of stars blink at us. (The inhabitants of the Southern Hemisphere are lucky enough to see twice as many as those in the Northern Hemisphere—they see a silvery rather than a black sky.) We also perceive among the stars wispy patches of clouds that are the billions of galaxies in the universe: One of them, the Milky Way, looks like a white gauzy river across our sky. Also visible are the *nebulae,*[7] clusters and superclusters of galaxies consisting of dust and gas clouds, within which stars were and are still born, each shining of its own light generated by nuclear fusion. These nebulae contain as many as 10,000 galaxies, some of

---

[6] Centrifugal force is analogous to the outward pull exerted on the hand spinning a ball on a string preventing the ball from flying out.

[7] From the Latin for clouds.

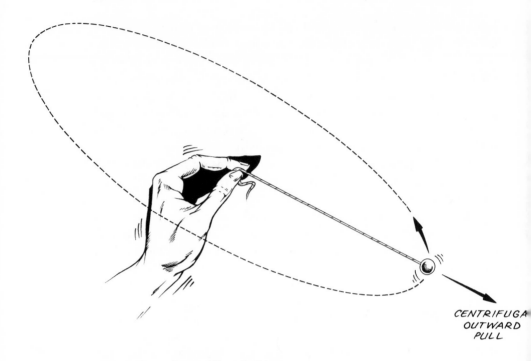

CENTRIFUGA
OUTWARD
PULL

Fig. 15.5   Centrifugal Force

irregular shape, some shaped like elliptic disks and others, like our own Milky Way, called *spiral galaxies*, in the shape of round disks with many curved arms (Fig. 15.6). The clusters are tens of millions of light years across the universe. Our *local group* contains many galaxies, of which the two largest, the Milky Way and M31 in the constellation Andromeda, have as many as 400 billion and 1,000 billion stars respectively. The distances to these galaxies are so huge that we cannot hope to set foot even on Andromeda, the closest, at only two million light years from earth, or for that matter, on any other body outside our own solar system. They are simply too incredibly far away from us mere mortals.

Our Still Expanding Universe

The rest of the story of the universe has lasted at least fourteen billion years, and appears to be a repetition of the first two billion: galaxies moving through the universe at inconceivable speeds until they disappear from our field of vision; stars being born, as they still are, shining for a few or many billions of years and then

Fig. 15.6   Spiral Galaxy

burning out (Fig. 15.7). But it is only in the last few years that we have reached an understanding of the universe, and a fuzzy one at that, because we are as yet unable to predict its end. That end, if there is ever going to be one, depends essentially on how much mass there is in the universe. If the density of matter in the universe is greater than a given critical value, the attractive force of gravity will slow down its expansion and the universe will end up in a Big Crunch, only to bounce back again in a new Big Bang. But, if the universe's density of mass is less than the critical value, the universe is bound to expand forever and, since the density of matter within it will get smaller and smaller and move outward faster and faster, as far as we human beings are concerned (and assuming that we will still be here), it will seem to disappear.

At present, astronomers are discovering increasing amounts of matter in the universe, some of it, in **black holes,** invisible concentrations of enormous masses of matter. It seems now that the density of matter may be at least five times greater than originally believed, most of it in masses of *dark matter* we are just beginning

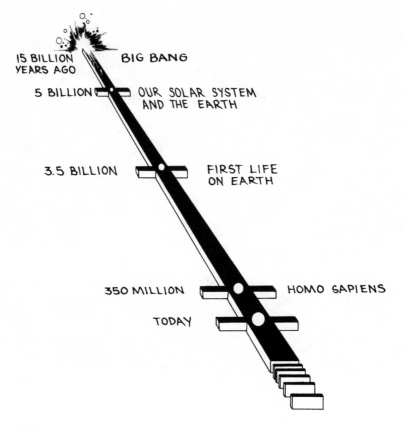

Fig. 15.7   Since the Big Bang

to understand. Moreover, we don't know whether ours is the only universe in some larger space of other universes. These questions have not been answered yet and are beyond the scope of this book: they are better left to the scientific speculations of the cosmologists. For now, the optimists who love the universe may be reassured: it may continue to "bang" and "crunch" forever.

Our Little Corner of the Universe

We are in the Milky Way galaxy, a minor member of the family of stars; our small corner in it, located in the constellation Orion, is a solar system that contains a single star, our sun (Fig. 15.8). A

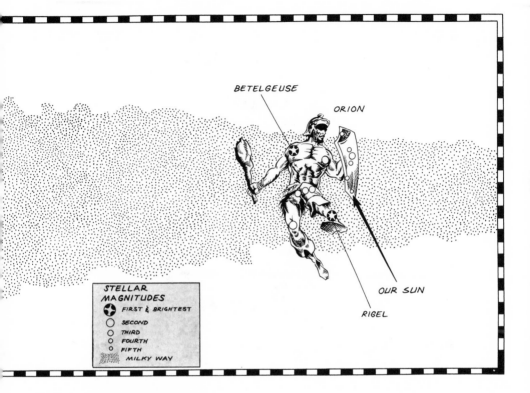

Fig. 15.8   Our Sun in Orion

small secondary star, assembled from the remnants of other stars' explosions rather than from the hydrogen and helium of the Big Bang, the sun is an incandescent ball of gases, 60 percent hydrogen by weight and 333,000 times the mass of the earth. Fusion reactions in its inner core (Fig. 15.9) generate enough heat and light radiating toward earth to sustain life on our planet. The sun took 800 million years to reach its present modest brightness (there are stars thousands of times brighter than the sun) and, by the standards of the universe, it has a modest temperature of 15 million degrees Kelvin in its core. We should not worry too much about our future, because the sun will continue radiating all the energy we need for another five billion years before it uses up its fusion fuel and begins to die by expanding into a huge, cool **red giant.** It will then collapse into a tiny, hot **white dwarf,** after a lifespan of ten billion years.

Some time long ago, our sun contracted and blew off the mat-

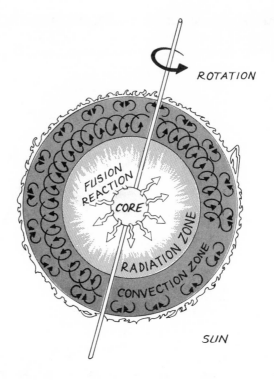

Fig. 15.9   The Sun

ter of its outer skin. Some of this matter eventually condensed into the planets of the solar system that spin around the sun. The solar system consists of five small planets (Mercury, Venus, Earth, Mars and the distant minor planet, Pluto) and four giant planets (Jupiter, Saturn, Uranus and Neptune), all circling the sun along elliptic *orbits*[8] lying on the same plane, a pretty good piece of evidence that they were all spun off by the sun (Fig. 15.10). We, the Earth, the third from the sun, are only 8.3 light minutes from it; Pluto, the farthest out, 5.5 light hours (to appreciate the miniscule dimensions of our system, you may compare them with the distance from us of the nearest star, Proxima Centauri, 4.28 light years away!)

The earth has its own little *satellite*,[9] the moon. It may have been spun from the earth as a result of a collision with another planetary object soon after the earth was formed, as evidenced by

[8] From the Latin word *orbita* for orbit.
[9] From the Latin word *satelles* for attendant.

Fig. 15.10   Our Solar System

the fact that the moon is 4.5 billion years old, almost as old as the earth, or it may have been formed by accretion from planetesimals (the small bodies rotating around the sun) at about the same time as the earth. Earth's moon, which circles the earth in about 29.5 days, is responsible for both the solar and the lunar eclipses, when it moves either between the earth and the sun or behind the earth (see Fig. 2). Its gravitational force slightly influences volcanoes and earthquakes by pulling on the crust of the earth, as well as the waters of our oceans.

When we compare our tiny corner of the universe to the infinite majesty of the whole, our innate, unjustified pride should naturally vanish. Yet, if we realize that, small as we are, we have been able to penetrate the mystery of the universe, establish its history, analyze its composition and draw its map; and that, moreover, we might even soon be able to predict its behavior to the end of time; and finally that, unfazed by our smallness, we have explored our little home, discovering in the process the origin of earthquakes and volcanoes, we are entitled to a complacent smile, a smile wisely tempered by the comparison of our mortality to the apparent never-ending life of the universe.

Glossary

Black hole. A region in space that is so dense and therefore has such strong gravitational force that even light cannot escape from it, although some radiation does.

Caldera. A large crater-like basin of a volcano that results from the explosion or collapse of the cone.

Dinosaur extinction. The discovery in 1992 of a buried 178 km (111 mi.) wide crater in the Yucatan Peninsula of Mexico, supports the theory that a 9 km (6 mi.) diameter meteorite struck the earth sixty-five million years ago. Such an impact would have melted and vaporized part of the earth's crust and thrown molten rock and deadly gasses into the air, leading to the extinction of major life forms. The 1994 impact of a 3.2 km- (2 mi.) diameter fragment of the Shoemaker-Levy comet with Jupiter provided us with an image of the effect of such a giant explosion.

Doppler effect and the **red shift.** First explained for sound waves
by the Austrian scientist Christian Doppler, this effect explains
the changes in wavelength, and hence in the frequency of any
type of wave, when the relative distance between the wave
source and the wave point of reception changes. For example,
it explains why the pitch of a locomotive whistle sounds higher
when moving toward us and lower when moving away (Fig.
G.1).

The Doppler effect explains the shift of the spectrum lines
toward the low frequency (red) end of the spectrum for a receding light source and toward the high frequency (blue) end for
an approaching light source (see Fig. 15.4).

Passing the light from a star through a spectroscope, we

Fig. G 1   The Doppler Effect

observe the lines that represent specific elements, such as hydrogen, shift toward the red from their expected location on the spectrum on earth. This "red shift" proves that the stars are receding from earth (and at a speed proportional to their distance from the earth).

Earth's Characteristics

Equatorial diameter	12 756 km (7,816 mi.)
Mass	6×10^{21} tonnes (5.4×10^{21} tons)
Density (water $= 1$)	5.52
Day	23 hr, 56 min, 4.09 sec
Equatorial gravity acceleration	9.78 m / sec^2 (32 ft. / sec.^{2})
Orbital velocity	29.8 km / sec (18.5 mi. / sec.)
Age	5 billion years

Einstein's hypotheses. Between 1905 and 1916 Albert Einstein (1879–1955) postulated four basic hypotheses that have revolutionized the field of physics. In elementary terms they state that:

1) Light is a flow of massless particles of energy, called *photons*, which behave both as particles and as waves because energy and mass are different aspects of the same physical entity (see 3). Since light consists of energy "particles", it is attracted by matter, as proved in the 1919 observation that the light from a star was bent by the gravitational pull of the sun, as could be seen during a solar eclipse (see Fig. 15.2).

2) The velocity of light, symbolized by the letter c, is a basic constant of nature and the greatest velocity possible in our universe ($c = 299\ 792$ km / sec or 186,282 mi. / sec.).

3) Energy, E, and mass, m, are equivalent aspects of the same physical entity. They are related by Einstein's equation:

$$E = mc^2,$$

which is the foundation of the theory of atomic energy used in building atomic reactors as well as nuclear bombs.

4) Gravity can be explained mathematically by adding time, t, to the coordinates x, y, z, of our three-dimensional space and treating these four dimensions on an equal footing. Gravity is measured by the curvature of a four-dimensional space.

Extrapolation. To extrapolate is to infer the value of a quantity outside the range where data exist, based on the trend of the known values of the quantity.

Four fundamental forces of nature.

a) The *gravitational force,* first postulated by Isaac Newton in 1665, is the attractive force between any two masses or amounts of matter. It is proportional to the product of the two masses divided by the square of the distance between them. The attraction between the earth's center and a mass is called the mass' *weight.* The range of the gravitational force is infinite.

b) The *electromagnetic force* is the force that affects the behavior of magnetic or charged particles. James Clerk Maxwell wrote the first theory of a unified magnetism and electricity in 1873. The range of electromagnetism is also infinite, and electromagnetic force is manifest in the generation of electricity.

c) The *weak nuclear force* is exerted between nuclear particles in reactions like the spontaneous decay of subatomic particles. It is stronger than the force of gravity but has the smallest range among the four universal forces. It is responsible for the energy released when fission or fusion occur, such as in the sun's interior.

d) The *strong nuclear force* acts between the smallest particles in the universe. These particles were first postulated by Murray Gell-Mann and George Zweig, who called them *quarks* (from a word invented by James Joyce in his last novel, *Finnegans Wake*). The strong nuclear force is the strongest of the four universal forces, and appears to be infinite in range, but is manifested only at microscopic scales because of its enormous strength. Quarks, in various combinations, constitute all the other nuclear particles in the universe. The strong force is responsible for binding quarks together, forming protons and neutrons. Apart from electrons and neutrinos, both called leptons, there are four kinds of quarks, one of which, the "top quark," has just been found experimentally in 1995.

Galaxy. A large grouping of stars containing possibly hundreds of billions of stars.

Inertia. The tendency of matter to remain at rest or, if moving, to continue moving in the same direction with the same speed unless affected by an outside force.

Isotopes. The isotopes of a given element behave chemically as that element but have different atomic weights: the nuclei of an isotope consist of the same number of protons, which determine its chemical behavior, but a variable number of neutrons, which determine its nuclear weight, different from that of the element.

Kelvin scale. The Kelvin scale (°K), uses the same interval as the Celsius scale, but has its zero at −273°C or *absolute zero*, the lowest temperature obtainable in the universe, at which all atoms and molecules stop moving. Zero on the Celsius scale is the freezing temperature of water and 100°C is the boiling point of water.

Lava. Melted rock (from underground magma) issuing from a fissure in the earth's crust.

Light spectrum. A white-light (continuous) spectrum is obtained by passing a light ray through a glass prism that splits it into a rainbow of colors with frequencies increasing from the low (infrared and red) end of the spectrum to the high (blue and ultraviolet) end (see Fig. 15.3).

  A given gas of any element will, at low pressure, emit only one line of a characteristic wavelength located at a specific frequency in the spectrum. By comparing the spectra of a star with the corresponding spectra for its elements on earth, scientists can evaluate the star's distance and speed away from earth.

Liquefaction occurs in sandy or silty soils lacking the binding property of clays. When such soils are completely saturated with water, the slightest shaking will separate the grains of the soil, allowing the water to act as a lubricant and giving the mixture a soupy consistency. Structures founded on such soils sink or tilt during an earthquake, as happened in 1964 to an apartment building in Nigata, Japan, which tilted through practically 90°, permitting the survivors to walk "down" the face of the building (Fig. G.2).

Fig. G 2   Nigata Apartment Buildings: After the Earth-quake.

Modified Mercalli intensity scale.* A standard since 1931 for grad-ing the intensity of earthquakes:

    I. Not felt, except by a very few under especially favorable circumstances.

    II. Felt only by a few persons at rest, especially on upper floors of buildings. Delicately suspended objects may swing.

    III. Felt quite noticeably indoors, especially on upper floors of buildings, but not recognized as an earthquake by many people. Standing motorcars may rock slightly. Vibration like passing truck. Duration estimated.

    IV. During the day felt indoors by many, outdoors by few. At night some awakened. Dishes, windows, and doors dis-turbed; walls make creaking sound. Sensation like heavy truck striking building. Standing motorcars rocked noticeably.

    V. Felt by nearly everyone; many awakened. Some dishes, windows etc., broken; a few instances of cracked plaster; unstable objects overturned. Disturbances of trees, poles and other tall objects sometimes noticed. Pendulum clocks may stop.

    *H. O. Wood and Frank Newman, *Bulletin of the Seismological Society of America*, 21, no. 4 (Dec. 1931).

VI. Felt by all; many frightened and run outdoors. Some heavy furniture moved; a few instances of fallen plaster or damaged chimneys. Damage slight.

VII. Everybody runs outdoors. Damage *negligible* in buildings of good design and construction; *slight to moderate* in well built ordinary structures; *considerable* in poorly built or badly designed structures. Some chimneys broken. Noticed by persons driving motorcars.

VIII. Damage *slight* in specially designated structures; *considerable* in ordinary substantial buildings, with partial collapse; *great* in poorly built structures. Panel walls thrown out of frame structures. Fall of chimneys, factory stacks, columns, monuments, walls. Heavy furniture overturned. Sand and mud ejected in small amounts. Changes in well water. Persons driving motorcars disturbed.

IX. Damage *considerable* in specially designed structures; well-designed frame structures thrown out of plumb; *great* in substantial buildings, with partial collapse. Buildings shifted off foundations. Ground cracked conspicuously. Underground pipes broken.

X. Some well-built wooden structures destroyed; most masonry and frame structures destroyed with foundations; ground badly cracked. Rails bent. Landslides considerable from river banks.

XI. Few, if any (masonry), structures remain standing. Bridges destroyed. Broad fissures in ground. Underground pipelines completely out of service. Earth slumps and land slips in soft ground. Rails bent greatly.

XII. Damage *total*. Waves seen on ground surfaces. Lines of sight and level distorted. Objects thrown upward into the air.

Neutron. Uncharged particle of slightly greater mass than a **proton** contained within the nucleus of an atom.

Newton's laws (more correctly, Newton's hypotheses).

1) A body will not change direction or speed unless acted upon by a force. (Astronauts, in the virtually gravity-free environment of space, move only when pushed by other astronauts or when they push or pull against the surfaces of their spaceship.)

2) A body changes speed—*accelerates*—at a rate proportional to the applied force. (At the takeoff of an airplane, we are increasingly pushed back against our seat as the plane accelerates forward, showing that the greater the acceleration, the greater the force and conversely, that the greater the force, the greater the acceleration.)

3) Two bodies behave *as if* they attract each other with a force proportional to the product of their masses and inversely proportional to the square of the distance between them. Since the force of gravity at the earth's surface is almost constant (because the earth is an almost perfect sphere), stepping on a scale will measure our weight to be the same anywhere on earth. On the moon, our weight is one-sixth that recorded on earth, because the moon's mass and therefore gravitational force is one sixth-that of the earth's.

Planck time. A small time interval such that 10 million of trillions of trillions of trillions of Planck times add up to a second. It is usually expressed in exponential notation as 10^{-43} seconds.

Proton. Positively charged elementary particle contained within the nucleus of an atom.

Pyroclastic flow. The flow of ash and mud in a volcanic eruption that may take one of three forms:

a) a glowing ash cloud, also called *nuée ardente*, with temperatures of up to 1 000°C (1,800°F) traveling at more than 100 km / hr (62 mph);

b) a cooler (and slower) ash flow;

c) a mud flow, consisting of cool ash and jumbled rock fragments mixed with water, that travels at less than 90 km / hr (54 mph).

Quantum theory. Developed by a number of scientists between 1900 and 1930, the basic hypotheses of quantum theory state that:

a) Energy is only generated in finite amounts called *quanta* but may also be generated in a continuous flow. (This hypothesis was first postulated by Max Plank in 1900.)

b) The behavior of individual particles cannot be predicted exactly and only their probability of being in a certain place and moving with a certain velocity, can be calculated.

c) According to the *principle of indeterminacy*, postulated by Werner Heisenberg in 1927, it is impossible to determine both the exact location and the exact velocity of a particle at a given time. Since the future course of a particle cannot be determined without knowing its position and velocity at a given time, quantum theory restricts the principle of *physical causality*. This controversial statement can be made clear by assuming that it is due to the unavoidable influence of the observer on the physical phenomenon being observed.

Radiocarbon dating. Carbon is present in most matter. By comparing the level of decay in a specimen of carbon 14, a radioisotope of carbon, against its standard rate of decay, the age of an archeological specimen or fossil can be determined.

Red giant. As a star runs out of fuel, its core shrinks and, as it burns helium, appears red, although the star's outer layers, which are still burning hydrogen, expand to a huge size. The star has the appearance of a giant red star because of the halo effect from the expanding outer layer.

Seiche. Damaging wave produced by the shaking of a closed body of water, such as a lake or reservoir (Fig. G.3).

Seismographs (seismometers). *Seismometers* are primitive instruments recording the intensity of earthquakes: they date from

Fig. G 3   Seiche Caused by an Earthquake

the Han Dynasty (202 B.C.–A.D. 220). *Seismographs* record earthquake intensities versus time. The first modern seismograph was invented by John Milne in 1885: it consisted of a pendulum, a heavy mass at the end of an arm that, because of its inertia, remains stationary while the stand (which holds the hinged pendulum firmly attached to the earth), moves with the earth (see Fig. 5.5). Modern seismographs use a coil of wire as the inertial mass; the coil is placed in a magnetic field so that a voltage is induced across it as the field is moved with respect to the coil. This electrical signal drives a stylus that records its trace on a *seismogram,* a paper-covered drum.

Stress and strain. *Stress* is the amount of force exerted on a unit area of a material and is usually expressed in kilograms per square millimeter, kg / mm² (pounds per square inch [psi]).

   The *strain* is the change in length of a unit length of material measured in mm / mm (in. / in.) due to a stress pulling or pushing on the material (Fig. G.4).

   When the stress acts in a direction parallel to the unit area of the material (a *shear stress*), the strain is the change in angle of an originally right angle. It is then called a *shear strain* (Fig. G.5).

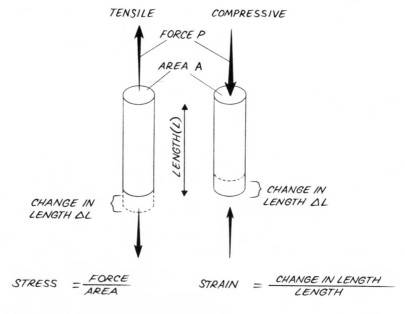

Fig. G 4   Axial Stress and Strain

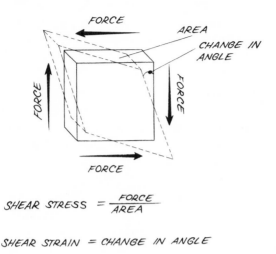

$$SHEAR\ STRESS\ =\ \frac{FORCE}{AREA}$$

$$SHEAR\ STRAIN\ =\ CHANGE\ IN\ ANGLE$$

Fig. G 5 Shear Stress and Strain

Supernova explosion. A delicate balance exists within a star between the outward pushing force of continual nuclear fusion and the inward pulling force of gravity. When this balance is upset by a reduced fusion reaction, the star becomes unstable and collapses on itself. The shock wave that moves outward from the collapsed star blows out the outer layers of the star with tremendous force, causing particles in its path to fuse together into new elements due to the released heat and energy of the explosion.

There are two types of supernovae: in the first, when a white dwarf grows to more than 1.4 times the mass of the sun, a chain of fusion reactions leads to an explosion; in the second, a star greater than 8 times the mass of the sun consumes its fuel in a series of fusion reactions, collapses under the force of its own gravity and finally blows apart in a spectacular event visible from earth.

Tsunami. From the Japanese words "tsu", meaning seashore village and "nami", meaning wave. These giant seawaves originate when the level of the sea bottom drops suddenly, usually along a tectonic boundary, due to an earthquake or a volcanic eruption. The sea pours in to fill the created void, then retreats as a wave. These waves have long *periods* (the time that passes from one wave crest to the next) ranging from five to over sixty minutes. The wave height in the open sea may be much less than a meter but, because of its long period, the volume of water between one crest and the next is so enormous that when it hits land it arrives as a crashing wall of water dozens of meters (over 30 ft) high. The crashing is mostly due to the bottom of the wave being slowed by friction with the ground while the top keeps traveling at high speed (see Fig. 4.3). Tsunamis travel at "airplane" speeds of up to 640 km / hr (400 mi / hr) and offer little time for warning before reaching land with devastating force. A system of four electronic warning stations is being deployed at the bottom of the Pacific Ocean (off the coast of Alaska and off the U.S. Pacific Northwest), to provide early warning of a tsunami's arrival. The stations will send signals of the size, speed and direction of the waves generated by the sudden motion of the ocean floor to buoys on the surface; the buoys will broadcast the information to stations in Alaska and Hawaii. The system, which is expected to be fully operative by the year 2000, should permit sufficient time for population evacuation and will thereby save lives.

Because of its size a tsunami can run up far inland of the seashore and when it withdraws will leave a wide area dry (drawdown).

White dwarf. It is created by the shrinking of a red giant to a size smaller than that of the earth, before slowly fading away as its core uses up its helium fuel. Some white dwarfs will pull in matter from nearby stars and, when reaching a mass greater than 140 percent of the sun's mass, will blow apart in a *supernova* explosion. It is the final stage in the life of a star.

Index